15

TESI
THESES

tesi di perfezionamento in Matematica sostenuta il 5 ottobre 2007

Paola Boito
Université de Limoges / CNRS
123 avenue Albert Thomas
87060 Limoges Cedex, France

Structured Matrix Based Methods for Approximate Polynomial GCD

Paola Boito

Structured Matrix Based Methods for Approximate Polynomial GCD

EDIZIONI
DELLA
NORMALE

ISBN: 978-88-7642-380-2
e-ISBN: 978-88-7642-381-9

"He found him under a pine tree, sitting on the ground, arranging fallen pine cones in a regular design: an isosceles triangle. At that hour of dawn Agilulf always needed to apply himself to some precise exercise: counting objects, arranging them in geometric patterns, resolving problems of arithmetic. It was the hour in which objects lose the consistency of shadow that accompanies them during the night and gradually reacquire colors, but seem to cross meanwhile an uncertain limbo, faintly touched, just breathed on by light; the hour in which one is least certain of the world's existence. He, Agilulf, always needed to feel himself facing things as if they were a massive wall against which he could pit the tension of his will, for only in this way did he manage to keep a sure consciousness of himself. But if the world around was instead melting into the vague and ambiguous, he would feel himself drowning in that morbid half light, incapable of allowing any clear thought or decision to flower in that void. In such moments he felt sick, faint; sometimes only at the cost of extreme effort did he feel himself able to avoid melting away completely. It was then he began to count: trees, leaves, stones, lances, pine cones, anything in front of him. Or he put them in rows and arranged them in squares and pyramids."

Italo Calvino, *The Nonexistent Knight*

Contents

Introduction

The computation of polynomial GCD is a basic algebraic task, which has many applications in several fields such as in polynomial root-finding, control theory, image deblurring, CAGD.

The problem is usually stated as follows: given the coefficients of two polynomials $u(x)$ and $v(x)$, compute the coefficients of their greatest common divisor $g(x)$. We focus here on the univariate case.

The usual notion of polynomial GCD, however, is ill-suited to deal with many applications where input data are affected by errors (due for example to roundoff, or to the fact that the data come from physical experiments or previous numerical computations). Indeed, if the given polynomials $u(x)$ and $v(x)$ have a nontrivial GCD, then arbitrarily small perturbations in the coefficients of $u(x)$ and $v(x)$ may transform $u(x)$ and $v(x)$ into relatively prime polynomials. Therefore the problem of finding an exact GCD is ill-posed in an approximate setting.

This is why the notion of approximate GCD has been introduced. Several definitions of approximate GCD are found in the literature; here we will mostly use the so-called ϵ-GCD. Roughly speaking, a polynomial $g(x)$ is an ϵ-GCD of $u(x)$ and $v(x)$ if there exist polynomials $\hat{u}(x)$ and $\hat{v}(x)$ such that

(i) $\hat{u}(x)$ and $\hat{v}(x)$ are "close" to $u(x)$ and $v(x)$, that is, $d(u, \hat{u}) < \epsilon$ and $d(v, \hat{v}) < \epsilon$ for some fixed polynomial metric d and tolerance ϵ;
(ii) $g(x)$ is an exact GCD of $\hat{u}(x)$ and $\hat{v}(x)$, and
(iii) $g(x)$ has maximum degree among the exact GCDs of pairs of polynomials that satisfy (i).

The first analysis of the approximate GCD problem dates back to 1985 ([118]); several approaches to the problem have been proposed since then. We seek here to give a comprehensive overview of the existing literature on the subject, with a focus on matrix-based methods. We next explore in detail the relationship between approximate GCD and resultant matrices (namely, Sylvester, Bézout and companion resultant matrices), their properties and factorizations. Three new algorithms for the computation of ϵ-GCD are presented:

- the algorithm TdBez uses Householder tridiagonalization of the Bézout matrix as its main tool;
- the algorithm PivQr is based on QR decomposition of the Sylvester matrix and subresultants, stabilized by column pivoting;
- the algorithm FastGcd exploits the Toeplitz-like structure of the Sylvester and Bézout matrices to compute an ϵ-GCD in a stable way and with a computational cost that is quadratic in the degrees of the input polynomials (whereas the complexity of known stable methods is at least cubic).

Chapters 1 to 5 present the definitions and formulations of the approximate GCD problem that can be found in the literature and outline the main ideas in the approaches proposed so far. Chapter 1 lists the definitions of quasi-GCD, ϵ-GCD, AGCD and δ-GCD and presents related topics that give useful insight into the approximate GCD problem, such as ϵ-root neighborhoods and a graph-theoretical description of approximate GCD.

Chapter 2 introduces the main tools that are necessary in a matrix-based approach to the polynomial GCD problem. Resultant matrices are defined and their relationship with polynomial GCD is examined. In particular, we prove some norm inequalities and a result on the QR decomposition of the Bézout matrix that will be useful later on. Resultant matrices have remarkable structure properties; therefore part of the chapter is devoted to a discussion of displacement structure and the fast method GKO for the factorization of structured matrices.

Chapter 3 analyses the use of variants of the Euclidean algorithm to compute an approximate GCD or estabilish approximate coprimeness. The basic version of the Euclidean algorithm has a low computational cost (*i.e.*, quadratic in the degree of the input polynomials), but it is numerically unstable. Several stabilized versions of the algorithm have been proposed; in some of them the crucial point is the choice of the termination criterion, whereas in other cases a look-ahead technique is employed to avoid ill-conditioned steps.

Chapter 4 is devoted to a description of known results and algorithms that rely on factorizations of the Sylvester matrix and subresultants. The singular value decomposition of the Sylvester matrix is often used to obtain estimates on the approximate rank of a resultant matrix, and therefore on the degree of an approximate GCD. The QR decomposition of the Sylvester matrix has been used in [41] to compute an approximate polynomial GCD, whereas the method outlined in [144] relies on the QR decomposition of Sylvester subresultants.

In most cases, algorithms for the computation of ϵ-GCD take a pair of polynomials and a tolerance ϵ as input, and output an ϵ-GCD. An alternative approach involves taking polynomials and the approximate GCD degree as input, and trying to minimize the norm of the perturbation that should be applied to the given polynomials so that they have an exact GCD of the prescribed degree. This is often called the optimization approach, and it is examined in Chapter 5.

Chapters 6 and 7 are devoted to the presentation of new methods for approximate GCD. The algorithms TdBez and PivQr are described in detail in Chapter 6, along with a study of the QR decomposition of resultant matrices and of the tridiagonalization of the Bezoutian. At the end of the chapter, three more algorithms for approximate GCD are briefly proposed. We feel that the algorithms described in this chapter, besides being quite effective, have the merit of highlighting some aspects of the interplay between resultant matrices and polynomial GCD that have been overlooked in the literature.

The main feature of the algorithm Fastgcd, presented in Chapter 7, is its low computational cost, combined with good stability properties. We show how a stabilized version of the GKO algorithm for the LU factorization of displacement structured matrices can be used to estimate the approximate GCD degree, compute the approximate GCD coefficients along with the associated cofactors, and perform iterative refinement.

The new algorithms presented here have been implemented in Matlab and applied to many test polynomials in order to evaluate the performance of these algorithms on typical "difficult" cases. Chapter 8 shows the most significant among these numerical experiments and compares the performance of our algorithms with the results given by other methods for which an implementation is available. A comparison between the notions of ϵ-GCD (based on perturbation of polynomial coefficients) and δ-GCD (based on perturbation of polynomial roots, see [103]) is also given in the last section.

Finally, Chapter 9 gives an overview of the many possible generalizations of the approximate GCD problem, as well as indications on further work.

Acknowledgements

First and foremost, I would like to thank my advisor, Dario Bini. I am grateful for his constant help and inspiring ideas, and for all the time and effort he devoted to the supervision of this thesis. Luca Gemignani

also gave a remarkable contribution to this work, both through useful suggestions and his thorough knowledge of the existing literature.

Part of the research work that led to this thesis was supported by the PRIN04 project "Structured matrix analysis: numerical methods and applications".

I would like to thank Victor Pan for suggesting the comparison between ϵ-GCD and δ-GCD in Section 8.10 and Zhonggang Zeng for sending me his software for approximate polynomial GCD.

I greatly appreciated the help that Filippo Callegaro, Giuseppe Della Sala, Antongiulio Fornasiero and Laura Luzzi gave me with technical issues. I also wish to mention Ivan Markowski, with whom I had stimulating discussions on the topic of polynomial GCD. My heartfelt thanks go to Roberto Grena, both for technical help and for his invaluable patience and support. Finally, I would like to thank my parents and all the people who have been close to me and have helped me throughout these years.

Pisa, 22 February 2007

Having defended my thesis, I would like to thank all the members of the committee and the referees, and in particular Bernhard Beckermann, who carefully examined this work and offered many insightful remarks. Most of the improvements in this revised version are due to his suggestions.

I would also like to thank Patrizia Gianni and Rob Corless for their helpful comments. I wish to mention Joab Winkler and John Allan: the summer school they organized in Oxford in September 2007 was for me a remarkable opportunity to learn about the latest developments on the subject of polynomial GCD and to meet people with similar interests.

Seeing my thesis published by Le Edizioni della Normale is a privilege I am very grateful for. My thanks go to the invaluable Luisa Ferrini, to Giuseppe Tomassini, to the referees and to all the editorial team.

Limoges, 22 April 2011

Notation

Fields of numbers are denoted here, as usual, by $\mathbb{R}$ (real) and $\mathbb{C}$ (complex). The imaginary unit is denoted by $\hat{\imath}$ (so as not to be confused with the letter i used as an index).

The vector space of real vectors of length n is denoted by $\mathbb{R}^n$ and the vector space of real $m \times n$ matrices is denoted by $\mathbb{R}^{m \times n}$. Analogously, $\mathbb{C}^n$ and $\mathbb{C}^{m \times n}$ are the vector spaces of complex n-vectors and $m \times n$ matrices, respectively.

We use capital letters (*e.g.* A, B, C) for matrices and boldface lower case letters for (column) vectors (*e.g.* **u**, **v**, **w**); the identity matrix of order n is denoted by I_n, whereas I is used when the size of the identity matrix is not explicitly specified. For diagonal matrices, the notation $A = \text{diag}(a_1, \ldots, a_n)$ stands for

$$A = \begin{pmatrix} a_1 & & O \\ & \ddots & \\ O & & a_n \end{pmatrix}.$$

Transpose matrices and vectors are denoted by A^T, $\mathbf{u}^T$, whereas the notation A^*, $\mathbf{u}^*$ is used to indicate transpose in the real case and Hermitian adjoint (*i.e.*, conjugate transpose) in the complex case.

Unless otherwise specified, a Matlab[1]-like notation is used for submatrices and matrix entries. Namely:

- $A(i, j)$ is the entry of A that belongs to the i-th row and to the j-th column;
- $A(i, :)$ is the i-th row of A and $A(:, j)$ is the j-th column of A;

---

[1] Matlab is a registered trademark of The MathWorks, Inc.

- $A(m{:}n, \; p{:}q)$ is the submatrix of A formed by the intersection of rows m to n and columns p to q.

Where explicitly stated, $\mathbf{u}$ denotes the column vector of length $n + 1$ associated with a univariate polynomial $u(x) = \sum_{i=0}^{n} u_i x^i$, i.e., $\mathbf{u} = [u_0, \, u_1, \, \ldots, u_n]^T$. See Section B.1 for notation in the multivariate case.

First-order approximations are sometimes denoted by $\doteq$ and $\lessdot$, so that $a \doteq b$ and $c \lessdot d$ mean $a = b + \mathcal{O}(\epsilon^2)$ and $c \leq d + \mathcal{O}(\epsilon^2)$ respectively.

Chapter 1
Approximate polynomial GCD

Finding the greatest common divisor (GCD) of two given polynomials is a basic problem in algebraic computing. The problem is usually stated as follows: given the (real or complex) coefficients of two polynomials, compute the coefficients of their greatest common divisor.

The range of applications is very wide; we mention here some examples.

- *Polynomial root-finding.* Computing the roots of a polynomial $p(x)$ which has multiple roots is an ill-conditioned problem. If a robust GCD finder is available, computing $g(x) = \text{GCD}(p, p')$ may help to solve this difficulty, since the roots of the polynomial $p(x)/g(x)$ will turn out to be better conditioned.
- *Simplifying rational functions.* Representing or performing computations with a rational function $R(x) = a(x)/b(x)$ might require $a(x)$ and $b(x)$ to be coprime. If $g(x) = \text{GCD}(a, b)$ is computed, then $R(x)$ can be replaced by $\tilde{R}(x) = \tilde{a}(x)/\tilde{b}(x)$, where $\tilde{a}(x) = a(x)/g(x)$ and $\tilde{b}(x) = b(x)/g(x)$. An application is degree reduction of rational curves, such as Bézier curves (see *e.g.* [120] and [18]).
- *Control theory.* Polynomial coprimeness is related to the controllability of linear control systems (see [7]).
- *Image restoration.* Polynomial GCD computations can be used for blind image deblurring (see [108]).

We will be mainly concerned here with the problem of evaluating the GCD of univariate polynomials $u(x)$ and $v(x)$.

The problem is well-understood in the exact case, that is, under the assumption that the coefficients of $u(x)$ and $v(x)$ are error-free. However, in many applications, input data are represented as floating point numbers or derive from the results of physical experiments or previous computation, so that they are generally affected by errors. The application of ordinary polynomial computations to such *empirical polynomials* is a field of study comprising elements from computer algebra and numerical analysis, to which a considerable amount of work has lately been devoted; see [121] for a review and further bibliography.

In our case, if $u(x)$ and $v(x)$ have a nontrivial GCD, it turns out that arbitrarily small perturbations in the coefficients of $u(x)$ and $v(x)$ may transform $u(x)$ and $v(x)$ into relatively prime polynomials. Therefore, it is clear that the concept of GCD is not well suited to deal with applications where data are approximatively known. This is why the notion of *approximate GCD* has been introduced.

1.1. Coefficient-based definitions

Starting from Schönhage ([118]), several different definitions of approximate polynomial GCD are found in the literature. The common underlying idea, however, is to look for a pair of polynomials $\hat{u}(x)$ and $\hat{v}(x)$ which are "close" to $u(x)$ and $v(x)$ and have a nontrivial exact GCD of maximum degree. The precise meaning of "close" depends both on the technical details of the definition and on the choice of a tolerance $\epsilon > 0$, which is related to the magnitude of the errors that may affect the coefficients of $u(x)$ and $v(x)$.

Throughout this work, the expression *approximate GCD* will be used to denote any of the different polynomials (quasi-GCD, ϵ-GCD, AGCD, δ-GCD) defined in the following sections, whereas the specific denomination will be used when appropriate. Notice that, while the acronym AGCD used by Zeng in [144] (see Section 1.1.3) actually stands for "approximate GCD", the abbreviated form will be reserved for Zeng's definition, so that the expression *approximate GCD* keeps its generic meaning.

1.1.1. Quasi-GCD

The first formalization of the notion of approximate GCD (*quasi-GCD*) is due to Schönhage ([118]) and dates back to 1985.

The definition is given at first for homogeneous polynomials, in order to account for the fact that a system of polynomial equations may have solutions close to (or at) infinity, which corresponds to nearly vanishing leading coefficients. Let

$$A(z_0, z_1) = \sum_{i=0}^{n} \alpha_i z_0^{n-i} z_1^i,$$

$$B(z_0, z_1) = \sum_{j=0}^{m} \beta_j z_0^{m-j} z_1^j,$$

with degrees $1 \leq n \leq m$. Define a polynomial norm as

$$|A| = \sum_{i=0}^{n} |\alpha_i|,$$

$$|B| = \sum_{i=0}^{m} |\beta_j|,$$

i.e. $|\cdot|$ is induced by the 1-norm applied to the vector of coefficients. It is also convenient to assume some kind of normalization on A and B, such as

$$|A|, |B| \in \left[\frac{1}{2}, 1\right].$$

Definition 1.1.1. Given $\epsilon > 0$, a homogeneous polynomial $H(z_0, z_1)$ of degree k is called a quasi-GCD of A and B within error ϵ if:

- there exist homogeneous polynomials A_1 of degree $n - k$ and B_1 of degree $m - k$ such that $|\Pi A_1 - A| < \epsilon$ and $|H B_1 - B| < \epsilon$;
- for any exact common divisor D of A and B there exists a homogeneous polynomial Q of degree $k - \deg D$ such that $|D Q - H| < \epsilon |H|$.

It is convenient, however, to reduce Definition 1.1.1 to the case of ordinary univariate polynomials. Let $f(z)$ and $g(z)$ be polynomials of degree n and m, respectively, with $m < n$, and let $0 < \epsilon \leq 1/2$. Let $\rho(f)$ be the root radius of $f(z)$, that is,

$$\rho(f) = \max_{i=1,\dots,n} \{|z_i| \quad \text{such} \quad \text{that} \quad f(z_i) = 0\}.$$

As a normalization condition, assume that $|f|, |g| \in [\frac{1}{2}, 1]$ and f has bounded root radius, *e.g.* $\rho(f) \leq 1/4$.

Definition 1.1.2. A polynomial $h(x)$ is a quasi-GCD for f and g with tolerance ϵ if there are polynomials $u(z)$ and $v(z)$ of degree $m - 1$ and $n - 1$ respectively, such that:

- $|h f_1 - f| < \epsilon$, $|h g_1 - g| < \epsilon$ for suitable f_1, g_1;
- $|u f + v g - h| < \epsilon |h|$.

The problem of quasi-GCD computation can therefore be stated as follows: given the coefficients of polynomials $f(z)$ and $g(z)$ and given ϵ as above, compute the coefficients of polynomials $h(z)$, $u(z)$ and $v(z)$ that satisfy Definition 1.1.2.

Schönhage proposes and discusses an algorithm for quasi-GCD computation, based on a modification of the Euclidean algorithm with pivoting.
This part of his pioneering work, while theoretically interesting, is of little use for practical purposes, because input numbers are assumed to be
available at any desired precision. In other words, if a number $\alpha \in \mathbb{R}$
belongs to the input set, then an *oracle* called with an arbitrary parameter
s will deliver a rational number a such that $|\alpha - a| < 2^{-s}$. This is hardly
the case in most practical applications, where the input polynomials are
known only to a limited accuracy, once and for all.

1.1.2. ϵ−GCD

We will present in this section the definition of approximate GCD that is
most widely used in the literature (see *e.g.* [40, 49, 103, 63]). This is also
the definition that is used in most cases throughout this work.

Definition 1.1.3. Let $u(x)$ and $v(x)$ be univariate (real or complex) poly-
nomials, with $n = \deg u(x)$ and $m = \deg v(x)$. Choose $\|\cdot\|$ a polynomial
norm (see Section A.3) and ϵ a positive real number. Then a polynomial
$g(x)$ is called

- an ϵ-*divisor* of $u(x)$ and $v(x)$ if there exist perturbed polynomials
  $\hat{u}(x)$ and $\hat{v}(x)$ such that

$$\deg \hat{u}(x) \leq n,$$
$$\deg \hat{v}(x) \leq m,$$
$$\|\hat{u}(x) - u(x)\| \leq \epsilon, \qquad (1.1.1)$$
$$\|\hat{v}(x) - v(x)\| \leq \epsilon \qquad (1.1.2)$$

  and $g(x)$ is an exact divisor of $\hat{u}(x)$ and $\hat{v}(x)$;
- an ϵ-*GCD* of $u(x)$ and $v(x)$ if it is an ϵ-divisor of maximum degree.

A few comments about this definition are needed.

First of all, notice that the definition requires to choose a polynomial
norm. A common choice is the 2-norm of the vector of coefficients, or another vector-induced norm; however, for some purposes one might want
to use a different norm, or even a polynomial distance not necessarily
induced by a norm. See Section A.3 for a brief discussion of this topic.

It should also be observed that several authors (*e.g.* [103]) prefer to
use a *normalized version* of Definition 1.1.3, replacing (1.1.1) and (1.1.2)
with

$$\|\hat{u}(x) - u(x)\| \leq \epsilon \|u(x)\|,$$
$$\|\hat{v}(x) - v(x)\| \leq \epsilon \|v(x)\|.$$

Lastly, it is important to point out that the ϵ-GCD, as defined here, is not unique. More precisely, its degree is uniquely defined, but its coefficients are not. This does not only happen because of the lack of normalization requirements on $g(x)$; there might be – and usually are – several polynomials that satisfy Definition 1.1.3, even without being scalar multiples of each other.

1.1.3. AGCD

In [144], Zhonggang Zeng points out that an approximate polynomial GCD (AGCD) for a set of polynomials should exhibit the following characteristics:

1. *nearness*: an AGCD is the exact GCD of a set of polynomials which are close to the given ones;
2. *maximum degree*: an AGCD has maximum degree among the polynomials that satisfy (1);
3. *minimum distance*: an AGCD minimizes the distance between the given set of polynomials and the set of polynomials of which it is the exact GCD.

Nearness and maximum degree are the key ideas shared by all the definitions of approximate GCD. Minimum distance is not always addressed in the literature, but it can certainly be desirable, though maybe difficult to achieve or check with certainty.

In order to achieve nearness, maximum degree and minimum distance, Zeng describes the AGCD problem as follows. Let $p_1(x), \ldots, p_l(x)$ be polynomials of degrees $m_1, \ldots, m_l$ respectively.

Saying that a polynomial $u(x)$ is an exact common divisor of fixed degree k for the p_i's means that there exist polynomials $v_1(x), \ldots, v_l(x)$ such that $p_i(x) = u(x)v_i(x)$ for all $i = 1, \ldots, l$. But these equations characterize $u(x)$ only up to multiplication by a constant; so one might want to add some normalization condition on $u(x)$, which can be expressed as $\mathbf{r}^*\mathbf{u} = 1$ for some given vector $\mathbf{r}$. For example, if $u(x)$ is expected to be monic, then $\mathbf{r}$ will be chosen as $[1 \quad 0 \quad \ldots \quad 0]^*$. So one obtains the following system:

$$F(\mathbf{z}) = b, \qquad (1.1.3)$$

where

$$F(\mathbf{z}) = \begin{bmatrix} \mathbf{r}^H\mathbf{u} - 1 \\ C_k(v_1)\mathbf{u} \\ \vdots \\ C_k(v_l)\mathbf{u} \end{bmatrix}, \quad \mathbf{z} = \begin{bmatrix} \mathbf{u} \\ \mathbf{v}_1 \\ \vdots \\ \mathbf{v}_l \end{bmatrix}, \quad b = \begin{bmatrix} 0 \\ \mathbf{p}_1 \\ \vdots \\ \mathbf{p}_l \end{bmatrix}$$

and each $C(v_i)$ is a convolution matrix associated with $v_i(x)$ (see Section B.1 for a definition).

In the approximate case, (1.1.3) is an overdetermined system; therefore a least squares solution is sought, so that the distance $\|F(\mathbf{z}) - \mathbf{b}\|_2$ is minimized.

Also notice the following fact.

Lemma 1.1.4. *If the distance $\|F(\mathbf{z}) - \mathbf{b}\|_2$ reaches a minimum at $\mathbf{z}$, then*

$$J(\mathbf{z})^*[F(\mathbf{z}) - \mathbf{b}] = 0, \tag{1.1.4}$$

where

$$J(\mathbf{z}) = \begin{bmatrix} \mathbf{r}^* & & \\ C_k(v_1) & C_{m_1-k}(u) & \\ \vdots & & \ddots \\ C_k(v_l) & & C_{m_l-k} \end{bmatrix}$$

is the Jacobian of $F(\mathbf{z})$.

The above discussion provides a motivation for the following definition.

Definition 1.1.5. The polynomial $u(x)$ is an AGCD of $p_1(x), \ldots, p_l(x)$ with tolerance ϵ if $u(x)$ is of the highest degree k, along with cofactors $v_j(x)$ of degree $m_j - k$, for $j = 1, \ldots, l$, such that:

- $\|F(\mathbf{z}) - \mathbf{b}\|_2 \leq \epsilon$,
- $J(\mathbf{z})^*[F(\mathbf{z}) - \mathbf{b}] = 0$.

Equation (1.1.4) provides a necessary (but not sufficient) condition for $\mathbf{z}$ to minimize $\|F(\mathbf{z}) - \mathbf{b}\|_2$. So it should be pointed out that an AGCD, according to Definition 1.1.5, is not guaranteed to satisfy the "minimum distance" requirement. Checking if $\mathbf{z}$ really minimizes the distance can be quite troublesome, which is why the AGCD definition settles for a compromise.

1.2. A geometrical interpretation

Emiris *et al.* outline in [50] an interesting geometrical approach to the notion of degree of an approximate polynomial GCD.

Let T_n be the set of monic univariate polynomials of degree n, with complex coefficients. A polynomial $f \in T_n$ can be denoted, using either its coefficients or its roots, as

$$f = x^n + \sum_{i=1}^{n} a_i x^{n-i} = \prod_{i=1}^{n} (x - \alpha_i).$$

Recall that the coefficients a_i are symmetric functions of the roots $\{\alpha_j\}$ and they are homogeneous polynomials of degree i. Also observe that f is defined by its roots, up to permutations. As is customary in the literature, we will denote with $\mathfrak{S}_n$ the group of permutations of n elements.

We therefore have the following diagram:

$$
\begin{array}{ccccc}
\mathbb{C}^n & \xrightarrow{\ \pi_n\ } & T_n = \mathbb{C}^n/\mathfrak{S}_n & \xrightarrow{\ \theta_n\ } & \mathbb{C}^n \\
(\alpha_1, \ldots, \alpha_n) & \longrightarrow & \{\alpha_1, \ldots, \alpha_n\} & \longrightarrow & (a_1, \ldots, a_n).
\end{array}
\tag{1.2.1}
$$

The projection π_n is surjective (it is actually a branched covering with $n!$ folds), whereas θ_n is a homeomorphism. It can be seen that T_n is endowed with the structure of a smooth analytical (and also algebraic) variety.

There are different ways to equip T_n with a metric. The usual distance induced by vector p-norms applied to the vectors of coefficients is obtained via θ_n. But it is also possible to define a distance using roots (i.e. via π_n): if a distance between sequences of roots has been chosen (using again, say, a vector p-norm), then set

$$
\mathrm{dist}(\{\alpha_j\}_j, \{\beta_j\}_j) = \min_{\sigma \in \mathfrak{S}_n} \mathrm{dist}((\alpha_j)_j, (\beta_{\sigma(j)})_j).
\tag{1.2.2}
$$

Now, let m and n be integers such that $m \leq n$. For ease of notation, we will write $f \in T_n$ rather that $\pi_n(f) \in T_n$ in what follows, thus identifying a monic polynomial with the set of its roots. For $r = 1, \ldots, m$, consider the following subsets of $T_n \times T_m$:

$$
\mathcal{G}_r = \{(f, g) \in T_n \times T_m : \deg(\mathrm{GCD}(f, g)) \geq r\}.
$$

It is easy to see that

$$
T_n \times T_m \supset \mathcal{G}_1 \supset \mathcal{G}_2 \supset \cdots \supset \mathcal{G}_m \simeq T_m.
$$

The following remarks give additional insight as to the role and properties of the set $\mathcal{G}_r$.

- Let S be the Sylvester matrix of f and g (see Section 2.5). As pointed out in (2.5.2), the condition $\deg(\mathrm{GCD}(f, g)) \geq r$ is equivalent to rank $S \leq m + n - r$, which in turn can be ensured by setting the appropriate minors of S equal to zero. Therefore $\mathcal{G}_r$ is an algebraic subvariety of $T_n \times T_m$ (identified with $\mathbb{C}^{m+n}$), and $\mathcal{G}_{r+1}$ is a closed subvariety of $\mathcal{G}_r$.

- Define a subvariety of $\mathbb{C}^{n+m}$ as the following union of $(n + m - r)$-planes:

$$S_r = \bigcup_{\sigma \in \mathfrak{S}_n, \tau \in \mathfrak{S}_m} \{\alpha_{\sigma(1)} = \beta_{\tau(1)}, \ldots, \alpha_{\sigma(r)} = \beta_{\tau(r)}\}.$$

Then $\mathcal{G}_r$ is the image of S_r via $\pi_n \times \pi_m$.

- Define the map

$$\begin{aligned} T_r \times T_{n-r} \times T_{m-r} &\longrightarrow T_n \times T_m \\ (p, f_r, g_r) &\longrightarrow (p \cdot f_r, p \cdot g_r). \end{aligned}$$

The image of this map is $\mathcal{G}_r$. Denote with ϕ_r the corresponding surjective map

$$\phi_r : T_r \times T_{n-r} \times T_{m-r} \longrightarrow \mathcal{G}_r.$$

Observe that ϕ_r is an isomorphism outside the preimage of $\mathcal{G}_{r+1}$.

Once a metric has been fixed on $T_n \times T_m$, the degree of an ϵ-GCD of a pair (f, g) can be defined as follows. Consider the set

$$\mathcal{G}_{r,\epsilon} = \bigcup_{(f,g) \in \mathcal{G}_r} B((f, g), \epsilon), \qquad r = 1, \ldots, m,$$

where $B((f, g), \epsilon)$ is the ball of center (f, g) and radius ϵ in $T_n \times T_m$, with respect to the chosen metric. Then

$$\deg(\epsilon\text{-GCD}(f, g)) \geq r \Leftrightarrow (f, g) \in \mathcal{G}_{r,\epsilon},$$

while the ϵ-GCD itself is an element of $\mathcal{G}_r \setminus \mathcal{G}_{r+1}$.

1.3. Pseudozeros and root neighborhoods

First proposed by Mosier ([98]), the ϵ-root neighborhoods of a polynomial are subsets of the complex plane whose properties provide some useful insight in relation to the problems of root-finding and detecting approximate polynomial coprimeness. The following presentation relies on [63, 98] and [127].

Let $\mathcal{P}_n$ be the set of polynomials with complex coefficients and degree at most n. Assign a metric d on $\mathcal{P}_n$ (see Section A.3 for a discussion of polynomial metrics; the weighted ∞-norm is the case originally studied by Mosier). An ϵ-neighborhood of $p(z)$ in $\mathcal{P}_n$ is the set

$$\mathcal{N}_\epsilon(p) = \{\hat{p} \in \mathcal{P}_n : d(p, \hat{p}) \leq \epsilon\}.$$

The ϵ-*pseudozero set*, or ϵ-root neighborhood, of $p(x)$ is then defined as the set of roots of polynomials belonging to an ϵ-neighborhood of $p(x)$:

$$\mathcal{Z}_\epsilon(p) = \{z \in \mathbb{C} : \exists \hat{p} \in \mathcal{N}_\epsilon(p), \ \hat{p}(z) = 0\}.$$

The connected components of $\mathcal{Z}_\epsilon(p)$ are called ϵ-*root neighborhoods* of $p(x)$.

Now, assume that $\mathcal{P}_n$ is equipped with a metric d induced by a norm $\|\cdot\|$ on $\mathbb{C}^{n+1}$ and let $\|\cdot\|_*$ denote the dual norm on $\mathbb{C}^{n+1}$. The vector $[1, z, \ldots, z^n]^T$ will be denoted as $\mathbf{z}$.

Define the function

$$g(z) = \frac{|p(z)|}{\|\mathbf{z}\|_*}.$$

A classical result (see *e.g.* [70]) states that there exists a vector $\mathbf{r} = [r_0, \ldots, r_n]^T \in \mathbb{C}^{n+1}$, called the *dual vector* of $\mathbf{z}$, such that $\|\mathbf{r}\| = 1$ and $\mathbf{r}^T \mathbf{z} = \|\mathbf{z}\|_*$. Choose $z \in \mathbb{C}$ and define

$$r(w) = \sum_{j=0}^{n} r_k w^k,$$

$$p_z(w) = p(w) - \frac{p(z)}{r(z)} r(w).$$

The following lemma says that $p_z(w)$ is the nearest to $p(w)$ among the polynomials having z as a root.

Lemma 1.3.1. *The polynomial $p_z(w)$ has the following properties:*

(i) $p_z(z) = 0$;
(ii) $p_z(w) \in \mathcal{N}_{g(z)}(p)$;
(iii) *if $q(w) \in \mathcal{P}_n$ and $q(z) = 0$, then $d(p, q) \geq d(p, p_z)$.*

Proof. The first property immediately follows from the definition of p_z. As for (ii), observe that

$$\|p_z - p\| = \frac{|p(z)|}{|r(z)|} \|r(w)\| = \frac{|p(z)|}{\mathbf{r}^T \mathbf{z}} \|\mathbf{r}\| = g(z).$$

In order to prove (iii), note that from Hölder's inequality it follows that

$$|p(w) - q(w)| \leq \|\mathbf{p} - \mathbf{q}\| \|\mathbf{w}\|_* = d(p, q) \|\mathbf{w}\|_*.$$

Since $q(z) = 0$, we have

$$|p(u)| \leq d(p, q) \|\mathbf{u}\|_*$$

and therefore

$$d(p, q) \leq g(z) = d(p, p_z).$$ ☐

We are now ready to prove the following theorem, which gives a characterization of ϵ-root neighborhoods in terms of the level curves of the function g.

Theorem 1.3.2.

$$\mathcal{Z}_\epsilon(p) = \{z \in \mathbb{C} : g(z) \leq \epsilon\}.$$

Proof. Let $z \in \mathcal{Z}_\epsilon(p)$; then by definition $\exists q \in \mathcal{N}_\epsilon(p)$ such that $q(z) = 0$ and $d(p, q) \leq \epsilon$. Then it follows from Lemma 1.3.1 that

$$|g(z)| = d(p, p_z) \leq d(p, q) \leq \epsilon.$$

Conversely, let $z \in \mathbb{C}$ be such that $g(z) \leq \epsilon$. Using Lemma 1.3.1 we have that $p_z(w)$ is a polynomial that belongs to $\mathcal{N}_\epsilon(p)$ and has z as a root, so $z \in \mathcal{Z}_\epsilon(p)$. ☐

Remark 1.3.3. The following particular cases of the above theorem are most frequently used in practical applications.

- If $d(p, q)$ is the distance induced by the relative 2-norm, *i.e.* $d(p,q) = \|\mathbf{p} - \mathbf{q}\|_2 / \|\mathbf{p}\|_2$, then

$$\mathcal{Z}_\epsilon(p) = \left\{z \in \mathbb{C} : \frac{|p(z)|}{\|\mathbf{z}\|_2 \|\mathbf{p}\|_2} \leq \epsilon\right\}. \tag{1.3.1}$$

- If $d(p,q)$ is the distance induced by the ∞-norm with weights given by the absolute values of the coefficients of p, *i.e.* $d(p,q) = \max_j \frac{|p_j - q_j|}{|p_j|}$, then

$$\mathcal{Z}_\epsilon(p) = \left\{z \in \mathbb{C} : \frac{|p(z)|}{\sum_{j=0}^n |p_j||z|^j} \leq \epsilon\right\}. \tag{1.3.2}$$

Theorem 1.3.2 and Remark 1.3.3 suggest a way to plot the ϵ-pseudozero set of a given polynomial $p(z)$, by computing the function $g(z)$ on a mesh and drawing the level curves $\{z \in \mathbb{C} : g(z) = \epsilon\}$. In Matlab this can be done using the `contour` command.

Figures 1.1, 1.2 and 1.3 show some examples of pseudozero sets computed in coefficient-wise relative ∞-norm, *i.e.* using (1.3.2).

It readily follows from the definition (or from Theorem 1.3.2) that the set $\mathcal{Z}_\epsilon(p)$ grows larger as ϵ increases. If ϵ is large enough, then $\mathcal{Z}_\epsilon(p)$

is no longer bounded. In the following we will assume that ϵ is small enough that the corresponding pseudozero set is bounded. This assumption requires no great restriction, since it essentially means that the polynomial $p(z)$ is known to be of degree n.

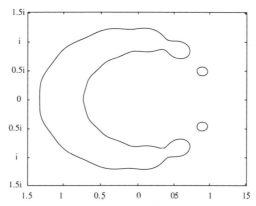

Figure 1.1. ϵ-pseudozero set of $p(z) = \sum_{j=0}^{12} z^j$ for $\epsilon = 0.15$.

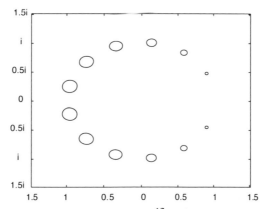

Figure 1.2. ϵ-pseudozero set of $p(z) = \sum_{j=0}^{12} z^j$ for $\epsilon = 0.05$.

The following result, which is found in [98], states more useful properties of the pseudozero set and explains the denomination of "ϵ-root neighborhoods".

Theorem 1.3.4. *Let* $q(z) \in \mathcal{N}_\epsilon(p)$. *Then* $p(z)$ *and* $q(z)$ *have the same number of roots, counting multiplicities, in each connected component of* $\mathcal{Z}_\epsilon(p)$. *Moreover, there is at least one root of* $p(z)$ *in each connected component of* $\mathcal{Z}_\epsilon(p)$.

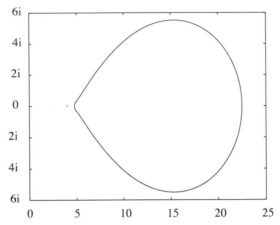

Figure 1.3. ϵ-pseudozero set of the Wilkinson polynomial $p(z) = \prod_{j=0}^{20}(x - j)$ for $\epsilon = 10^{-10}$.

Proof. It follows from the discussion preceding the theorem that both $p(z)$ and $q(z)$ are of degree n and therefore have n roots each, counting multiplicities. The same is true for every member of the family of polynomials

$$\mathcal{F} = f_\lambda(z) = \lambda p(z) + (1 - \lambda)q(z), \quad 0 \leq \lambda \leq 1.$$

Furthermore, all the roots of $f_\lambda(z)$, $0 \leq \lambda \leq 1$, belong to $\mathcal{Z}_{epsilon}(p)$. Indeed we have:

$$d(f_\lambda, p) = (1 - \lambda)d(q, p) \leq \epsilon.$$

Observe that the coefficients of $f_\lambda(z)$ are linear functions of λ; besides, it is a well-known fact that the roots of a polynomial are continuous functions of the coefficients. It follows that, as λ varies from 0 to 1, the roots of $f_\lambda(z)$ trace continuous paths in the complex plane, from the roots of $f_0(z) = q(z)$ to the roots of $f_1(z) = p(z)$. If $f_0(z)$ has k roots in some connected component of $\mathcal{Z}_\epsilon(p)$, while the remaining $n - k$ roots belong to the other connected components, the same will be true for every polynomial in $\mathcal{F}$. This proves the first assertion of the theorem.

   The second assertion follows from the first and from the definition of $\mathcal{Z}_\epsilon(p)$.   □

   As it can be seen from the definition and the previously discussed properties, pseudozeros are related to the conditioning of the root-finding problem. Ill-conditioned roots typically have large ϵ-neighborhood, as shown in Figure 1.3 for the case of the Wilkinson polynomial.

Let us now explore the connection between the pseudozero set and approximate polynomial primality. The following theorem is found in [63].

Theorem 1.3.5. *Let* $p(z) \in \mathcal{P}_n$, $q(z) \in \mathcal{P}_m$ *and* $\epsilon > 0$. *Then* $p(z)$ *and* $q(z)$ *are* ϵ-*coprime if and only if* $Z_\epsilon(p) \cap Z_\epsilon(q) = \emptyset$.

Proof. If $Z_\epsilon(p) \cap Z_\epsilon(q) = \emptyset$, then by definition of the pseudozero set there is no pair of polynomials $\hat{p}(z) \in \mathcal{N}_\epsilon(p)$, $\hat{q} \in \mathcal{N}_\epsilon(q)$ having a common root. Therefore, $p(z)$ and $q(z)$ are coprime.

Conversely, suppose $Z_\epsilon(p) \cap Z_\epsilon(q) \neq \emptyset$. Then we can choose $a \in Z_\epsilon(p) \cap Z_\epsilon(q)$. By definition, there are polynomials $\hat{p}(z) \in \mathcal{N}_\epsilon(p)$ and $\hat{q} \in \mathcal{N}_\epsilon(q)$ such that $\hat{p}(a) = 0$ and $\hat{q}(a) = 0$. But this means that the polynomial $(z - a)$ divides both $\hat{p}(z)$ and $\hat{q}(z)$, and therefore it is an ϵ-divisor of both $p(z)$ and $q(z)$.   □

Example 1.3.6. Let

$$p(z) = (x - 1)(x - 1.1)(x - (2 + 0.2i))(x - (2 - 0.2i))$$

$$= x^4 - 6.01x^3 + 13.09x^2 - 12.16x + 4.0804,$$

$$q(z) = (x - (0.8 + 0.05i))(x - (0.8 - 0.05i))(x - 0.5)^2$$

$$= x^4 - 2.6x^3 + 2.4925x^2 - 1.0425x + 0.160625.$$

Figures 1.4 and 1.5 show the ϵ-pseudozero sets of $p(z)$ (solid line) and $q(z)$ (dotted line) computed in relative 2-norm for different values of ϵ. It follows from Theorem 1.3.5 that $p(z)$ and $q(z)$ are ϵ-coprime for $\epsilon = 0.0001$ but not for $\epsilon = 0.0003$.

1.4. A root-based definition

It is also possible to define a notion of approximate polynomial GCD in terms of polynomial roots rather than coefficients. As pointed out in [103, 104, 105], the ϵ-GCD as defined in the previous sections may sometimes show an unexpected behaviour, particularly when the polynomials involved have multiple or clustered zeros.

Example 1.4.1. Let n be an even (and large) integer and define

$$p(x) = x^n - 2^{-n},$$
$$q(x) = (x + 1)^n.$$

The polynomial $q(x)$ has only one zero, namely -1, with multiplicity n. The zeros of $p(x)$ are $\{x_k = \frac{1}{2}e^{2\pi i/n}\}_{k=0,\dots,n-1}$. The distance between

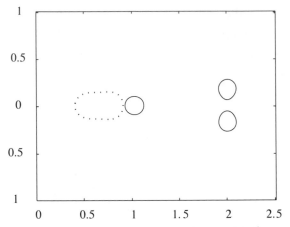

Figure 1.4. ϵ-pseudozero sets of $p(z)$ and $q(z)$ for $\epsilon = 10^{-4}$.

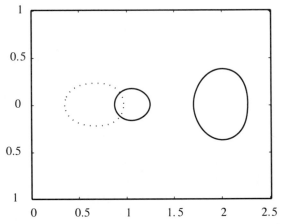

Figure 1.5. ϵ-pseudozero sets of $p(z)$ and $q(z)$ for $\epsilon = 3 \times 10^{-4}$.

the zeros of $p(x)$ and $q(x)$ is therefore quite large, and it is clear that $GCD(p, q) = 1$.

One would expect $p(x)$ and $q(x)$ to be coprime also from an approximate point of view. But it is easily verified that $\deg \epsilon\text{-}GCD(p, q) \geq 1$ for $\epsilon \geq 2^{-n}$; in other words, $p(x)$ and $q(x)$ may have a nontrivial ϵ-GCD even for small values of ϵ. See Figure 1.6.

Example 1.4.2. Another classical example involves polynomials

$$p(x) = x^n,$$
$$q(x) = (x - 1)^n.$$

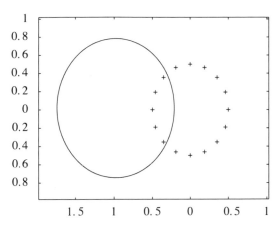

Figure 1.6. The curve on the left shows the ϵ-pseudozero set of $q(x)$ for $\epsilon = 10^{-8}$, with respect to the ∞-norm, whereas the "+" marks indicate the roots of $p(z)$, where $p(x)$ and $q(x)$ are as in Example 1.4.1 with $n = 16$.

Again, the roots of $p(x)$ and $q(x)$ are well separated; nonetheless, for every $\epsilon > 0$ there exists an integer n_0 such that $p(x)$ and $q(x)$ have a nontrivial ϵ-GCD whenever $n > n_0$.

The behaviour described above is explained by the ill-conditioning of the roots of the polynomials as functions of their coefficients, due to multiple or clustered roots. But it can be surprising if one expects the approximate GCD to closely resemble the exact GCD.

Therefore Example 1.4.1 provides the motivation to propose a definition of approximate polynomial GCD which is based on the perturbation of roots rather than coefficients, as has been done by V. Y. Pan in ([103, 104, 105]). This new notion of approximate GCD is denoted as δ-GCD to distinguish it from the ϵ-GCD based on perturbation of coefficients.

Let the polynomials $p(z)$ and $q(z)$ be defined through their roots as

$$p(z) = p_n \prod_{j=1}^{n} (z - \alpha_j),$$

$$q(z) = q_m \prod_{j=1}^{m} (z - \beta_j)$$

with $p_n q_m \neq 0$. The norm of a polynomial will also be defined as a vector norm of its roots (the ∞-norm will be used in the following).

The δ-neighborhoods of $p(x)$ and $q(x)$ are

$$\mathcal{N}_\delta(p) = \left\{ \hat{p}(z) = p_n \prod_{j=1}^{n} (z - \hat{\alpha}_j) : |\alpha_j - \hat{\alpha}_j| \leq \delta, \ j = 1, \ldots, n \right\},$$

$$\mathcal{N}_\delta(q) = \left\{ \hat{q}(z) = q_m \prod_{j=1}^{m} (z - \hat{\beta}_j) : |\beta_j - \hat{\beta}_j| \leq \delta, \ j = 1, \ldots, m \right\}.$$

A δ-*divisor* of $p(x)$ and $q(x)$ is a monic polynomial $g(z)$ that divides exactly some pair of polynomials $\hat{p} \in \mathcal{N}_\delta(p)$, $\hat{q} \in \mathcal{N}_\delta(q)$. A δ-divisor $g(z)$ of maximum degree d_δ is called δ-*GCD* of $p(z)$ and $q(z)$. Again, d_δ is uniquely defined but the δ-GCD is not.

Though the present work focuses on computational techniques for the ϵ-GCD rather then the δ-GCD, a method proposed by Pan for the study and computation of the δ-GCD is described in the next section. Moreover, Chapter 8 includes examples of computation of δ-GCDs and a comparison with ϵ-GCDs.

We point out here that Pan's definition of δ-GCD basically relies on a polynomial matric derived from the ∞-norm defined on the vectors of roots. An alternative approach, which may prove interesting in the case of very large roots, uses the stereographic distance on $\mathbb{C}$; see [86] for a discussion in the context of approximate partial fraction decomposition in $\mathbb{C}[z]$.

1.5. Graph-theoretical techniques for bounding the degree of the approximate GCDs

In [103], V. Y. Pan proposes a graph-theoretical approach to the computation of upper and lower bounds for the degree of an approximate GCD.
Let

$$p(x) = \sum_{i=0}^{n} p_i x^i = p_n \prod_{j=1}^{n} (x - \alpha_j).$$

For every $j = 1, \ldots, n$ define the ϵ-*perturbation domain* $D_{\epsilon,j}(p)$ associated with the zero α_j of $p(x)$ as the set formed by the images of α_j under all homotopic transformations of $p(x)$ in $\mathcal{N}_\epsilon(p)$. More precisely:

$$D_{\epsilon,j} = \{\tilde{\alpha} : \exists \tilde{p} \in \mathcal{N}_\delta(p) \quad \text{and} \quad \alpha(t) : [0,1] \longrightarrow \mathbb{C} \quad \text{continuous}$$
$$\text{such that} \quad t\tilde{p}(\alpha(t)) + (1-t)p(\alpha(t)) = 0 \quad \text{for} \quad 0 \leq t \leq 1,$$
$$\alpha_j = \alpha(0), \quad \tilde{\alpha} = \alpha(1)\}.$$

Observe that the union of ϵ-perturbation domains actually coincides with ϵ-pseudozero set (as defined in Section 1.3); compare the proof of Theorem 1.3.4.

However, ϵ-perturbation domains provide neighborhoods for each separate root, whereas each connected component of the set of ϵ-root neighborhoods may contain a cluster of roots.

Perturbation domains are usually difficult to compute explicitly; but they can be approximated using disc-neighborhoods of the roots. For every $j = 0, \ldots, n$ define discs $D_{\epsilon,j}^+(p)$ and $D_{\epsilon,j}^-(p)$ in the complex plane, such that

$$D_{\epsilon,j}^-(p) \subseteq D_{\epsilon,j}(p) \subseteq D_{\epsilon,j}^+(p).$$

Of course it is desirable that these discs should be as close to $D_{\epsilon,j}(p)$ as possible. Generally speaking, one may determine suitable discs by

1. approximating all the roots $\alpha_1, \ldots, \alpha_n$ within some fixed error bound (which is assumed to be small in comparison to ϵ), and
2. applying known estimates that relate perturbations on the coefficients of $p(x)$ to perturbations on the roots.

More precisely, the following results can be used to compute radii for $D_{\epsilon,j}^+(p)$ and $D_{\epsilon,j}^-(p)$.

Proposition 1.5.1. ([101, 117]) *Let $p(x)$ be such that $|\alpha_j| \le 1$ for all $j = 1, \ldots, n$. A perturbation of the coefficients of $p(x)$ within ϵ perturbs the zeros $\{\alpha_i\}$ by at most $4\epsilon^{1/k}$.*

Proposition 1.5.2. *If the zeros of a polynomial $p(x)$ are perturbed within a positive δ, with*

$$\delta \le (1 + \epsilon)^{1/k} - 1,$$

then the corresponding polynomial remains in $\mathcal{N}_\epsilon(p)$.

This procedure will later be described in more detail for the case of δ-GCD (see Section 1.4).

Now, given polynomials $p(x)$ and $q(x)$ of degrees n and m respectively, define three bipartite graphs G_ϵ^-, G_ϵ and G_ϵ^+. Recall that:

- a *bipartite graph* is a graph whose set of vertices are decomposed into two disjoint sets, such that no two vertices within the same set are connected by an edge;
- a *matching* is a set of edges in a graph such that no two of them share a vertex in common;
- a *clique* is a subgraph in which each pair of vertices is connected by an edge.

The graph $G_{\epsilon,j}^-$ is defined so as to have two sets of vertices $U^- = \{u_i^-\}$ and $V^- = \{v_i^-\}$, where each vertex u_i^- represents a disc $D_{\epsilon,j}^-(p)$ and each v_i^- represents a disc $D_{\epsilon,j}^-(q)$. A pair of vertices u_i^- and v_j^- are connected by an edge if and only if the discs $D_{\epsilon,i}^-(p)$ and $D_{\epsilon,j}^-(q)$ have non-empty intersection. The graphs G_ϵ and G_ϵ^+ are similarly defined, using of course perturbation domains and outer discs instead of inner discs. The following result is easily verified:

Proposition 1.5.3. *Let $|M^-|$, $|M|$ and $|M^+|$ be the cardinalities of maximum matchings in the bipartite graphs G_ϵ^-, G_ϵ and G_ϵ^+, respectively. Let d_ϵ be the degree of an ϵ-GCD of $p(x)$ and $q(x)$. Then $|M^-| \leq d_\epsilon = |M| \leq |M^+|$.*

In other words, $d^- = |M^-|$ and $d+ = |M^+|$ give upper and lower bounds on the degree of an ϵ-GCD. If $d^- = d^+$, a tentative ϵ-GCD may be computed as a polynomial whose roots are chosen in the intersection domains $D_{\epsilon,j}^-(p) \cap D_{\epsilon,j}^-(q)$. Techniques for computing maximum matchings can be found in [71] and [56].

The approach to the computation of an approximate GCD described so far is particularly well-suited for δ-GCDs, *i.e.* approximate GCDs defined through perturbation of the roots, rather than coefficients, of the polynomials involved. Indeed, in the case of δ-GCDs, perturbation domains are themselves disc-neighborhoods of the roots, with known radii; so there is no need to define discs $\{D_{\delta,j}^-\}$ and $\{D_{\delta,j}^-\}$ and the corresponding graphs. It is sufficient to work with perturbation domains $\{D_{\delta,j}\}$, which are in this case discs of radius δ and center the roots $\{\alpha_j\}$, and the associated bipartite graph G. The vertices $\{u_i\}$ represent the roots $\{\alpha_i\}$ of $p(x)$, whereas vertices $\{v_j\}$ represent the roots $\{\beta_j\}$ of $q(x)$. Two vertices u_k and v_h are connected by an edge if and only if $|\alpha_k - \beta_h| < 2\delta$. A maximum matching $M = \{(u_{i(1)}, v_{j(1)}), \ldots, (u_{i(d)}, v_{j(d)})\}$ immediately defines a certified δ-GCD of $p(x)$ and $q(x)$, which can be computed for example as

$$g(x) = \prod_{s=1}^{d} \left(x - \frac{\alpha_{i(s)} + \beta_{j(s)}}{2} \right).$$

The computational effort required for this algorithm involves $\mathcal{O}(n^2)$ ops and comparisons to construct the graph G and $\mathcal{O}(n^{2.5})$ comparisons to compute a maximum matching. The computation of the coefficients of $g(x)$, if necessary, can be done in $\mathcal{O}(n \log^2 n)$ ops (see [19]).

The computation of a maximum matching in G can be simplified if δ is allowed to increase dynamically.

First observe that a maximum matching is easily found if all the components in G are bipartite cliques. Also notice that there are at most $2n - 1$ edges in any simple path in G, and each edge connects two points u_k and v_h whose associated polynomial roots are such that $|\alpha_k - \beta_h| < 2\delta$. If δ is replaced with $\delta_1 = (4n - 2)\delta$ and the graph G is redefined accordingly, obtaining a new graph G_1, then every component of G becomes a bipartite clique in G_1.

If now all the components of G_1 are bipartite cliques, it is easy to compute a maximum matching and therefore a δ_1-GCD. Otherwise, the procedure is recursively repeated, replacing δ_{r-1} and G_{r-1} by $\delta_r = (4n - 2)^r \delta$ and G_r, until a graph whose components are all bipartite cliques is reached. The computational cost becomes $\mathcal{O}(mn)$.

1.6. Formulations of the approximate GCD problem

Once a definition of approximate GCD has been chosen, and univariate polynomials $u(x)$ and $v(x)$ have been assigned, there are several approaches to the study of approximate GCDs for $u(x)$ and $v(x)$, both from the theoretical and the computational point of view.

- The problem we will usually be concerned with can be formulated as follows:

  Problem 1.6.1. Given a tolerance ϵ and the (real or complex) coefficients of polynomials $u(x)$ and $v(x)$, compute the coefficients of an ϵ-GCD for $u(x)$ and $v(x)$ according to Definition 1.1.3, with or without normalization.

- A more basic question involves the notion of ϵ-coprime polynomials.

  Definition 1.6.2. The polynomials $u(x)$ and $v(x)$ are *approximately coprime* if their approximate GCD is trivial; in particular, if a tolerance ϵ has been chosen, then $u(x)$ and $v(x)$ are said to be ϵ-coprime if their ϵ-GCD is trivial.

  The problem then becomes:

  Problem 1.6.3. Given a tolerance ϵ and the (real or complex) coefficients of polynomials $u(x)$ and $v(x)$, estabilish whether $u(x)$ and $v(x)$ are ϵ-coprime.

- What is the minimum value for ϵ such that $u(x)$ and $v(x)$ have a non-trivial ϵ-GCD? This is clearly related to ϵ-primality.
- More generally, given an integer k, what is the minimum ϵ such that $u(x)$ and $v(x)$ have an ϵ-GCD of degree exactly/at least k? (See also Chapter 5 and Problem 5.0.2).

- Compute, if possible, a "natural" approximate GCD of $u(x)$ and $v(x)$ (the notion of "natural" degree of an approximate GCD will be explained in more detail in Section 4.1.1).
- Solve the above problems through a root-based approach: given a tolerance δ and the roots of polynomials $u(x)$ and $v(x)$, compute a δ-GCD for $u(x)$ and $v(x)$, or determine whether $u(x)$ and $v(x)$ are δ-coprime, etc.

Chapter 2
Structured and resultant matrices

A resultant matrix of two polynomials is a matrix whose entries are functions of the polynomial coefficients, and such that the polynomials have a
common root if and only if the determinant of this matrix (which is often
called the *resultant*) is zero. Moreover, the degree of the GCD of the two
polynomials is equal to the rank deficiency of the resultant matrix.

The Sylvester and Bézout matrices – which will be discussed here in
some detail – are well known examples of resultant matrices. They also
display a Toeplitz-block (in the Sylvester case) and a displacement structure; this is why numerical methods originally developed for the solution
of structured systems may prove useful when studying the polynomial
GCD problem.

2.1. Toeplitz and Hankel matrices

We will give in this section a brief presentation of a class of dense structured matrices, defined by linear constraints on the matrix entries. A more
general definition will be given in the next section.

Definition 2.1.1. $T = [t_{i,j}]$ is a *Toeplitz matrix* if $t_{i,j} = t_{i+k,j+k}$ for all
positive k, that is, if the entries of T are invariant with respect to shift
along the diagonal direction. A Toeplitz matrix is therefore completely
defined by its first row and first column.

Definition 2.1.2. $H = h_{i,j}$ is a *Hankel matrix* if $h_{i,j} = h_{i-k,j+k}$ for all
k, that is, if the entries of H are invariant with respect to shift along the
antidiagonal direction. A Hankel matrix is completely defined by its first
row and last column.

Remark 2.1.3. Define the permutation matrix

$$\begin{pmatrix} 0 & & & 0 & 1 \\ & & \cdot^{\cdot} & \cdot^{\cdot} & 0 \\ 0 & & \cdot^{\cdot} & \cdot^{\cdot} & \\ 1 & 0 & & & 0 \end{pmatrix}. \tag{2.1.1}$$

Then

- JT and TJ are Hankel matrices for any Toeplitz matrix T;
- JH and HJ are Toeplitz matrices for any Hankel matrix H.

An $n \times n$ Toeplitz or Hankel matrix is completely specified by $2n - 1$ parameters, thus requiring less storage space than ordinary dense matrices. Moreover, many computations with Toeplitz or Hankel matrices can be performed faster; this is the case, for instance, for the sum and the product by a scalar. Less trivial examples are given by the following results (see [19]):

Proposition 2.1.4. *The multiplication of an $n \times n$ Hankel or Toeplitz matrix by a vector can be reduced to multiplication of two polynomials of degree at most $2n$ and performed with a computational cost of $\mathcal{O}(n \log n)$.*

Proposition 2.1.5. *A nonsingular linear system of n equations with Hankel or Toeplitz matrix can be solved with a computational cost of $\mathcal{O}(n \log^2 n)$.*

It should be pointed out, however, that a low computational cost might come at the expense of stability. For a discussion of numerical stability of solving Toeplitz and Hankel linear system, see [42, 27, 92]. Fast methods for solving Toeplitz or Hankel systems which show better stability properties include direct methods with a quadratic cost (see *e.g.* [60], or below in this chapter) and iterative methods with a computational cost of $\mathcal{O}(n \log n)$ at each step ([93]).

2.2. Displacement structure

Following [79], let us first introduce the concept of displacement structure for the special case of a symmetric (or Hermitian) Toeplitz matrix T.

It has already been pointed out that many matrix problems involving T can be solved with complexity $\mathcal{O}(n^2)$ rather than $\mathcal{O}(n^3)$. This may not be surprising, given that, while generic square matrices depend on n^2 parameters, T only depends on n.

Nothing has been said, though, about inverses and products of Toeplitz matrices and related combinations; such matrices are not Toeplitz, but we might nonetheless expect them to display some kind of structure, which in turn might help to reduce the complexity of problems as inversion, factorization etc.

It turns out that the appropriate common property of all these matrices is, rather than their "Toeplitzness", their low *displacement rank*. Let us now explain in more detail what this means.

The displacement of an $n \times n$ Hermitian matrix R was originally defined in [77] as follows. Let Z be the $n \times n$ lower shift matrix, *i.e.*

$$Z = \begin{pmatrix} 0 & \cdots & \cdots & \cdots & 0 \\ 1 & 0 & \cdots & \cdots & \vdots \\ 0 & 1 & \ddots & & \vdots \\ \vdots & & \ddots & \ddots & \vdots \\ 0 & \cdots & 0 & 1 & 0 \end{pmatrix}.$$

Then the displacement of R is defined as

$$\nabla R = R - ZRZ^*. \tag{2.2.1}$$

Observe that ZRZ^* is obtained by shifting R downwards along the main diagonal; this is the reason for the choice of the name *displacement*.

If ∇R has low rank r, independent of n, then R is said to be *structured* with respect to the displacement operator (2.2.1), and r is called the *displacement rank* of R.

Notice that ∇R is also Hermitian. Therefore its eigenvalues are real and we can define the *displacement inertia* of R as the pair $\{p, q\}$, where p is the number of positive eigenvalues and q is the number of negative eigenvalues of ∇R; the sum of p and q is equal to the displacement rank. So we can write

$$\nabla R = R - ZRZ^* = GJG^*, \tag{2.2.2}$$

where

$$J = \begin{pmatrix} \mathbf{I}_p & \mathbf{0} \\ \mathbf{0} & -\mathbf{I}_q \end{pmatrix}$$

and G is an $n \times r$ matrix. The pair $\{G, J\}$ is called a *generator* of R and it contains all the information on R. The choice of G is not unique: for example, $G\Theta$ is also a generator for every Θ such that $\Theta J \Theta^* = J$.

Example 2.2.1. A symmetric Toeplitz matrix $T = [c_{|i-j|}]_{i,j=0}^{n}-1$, $c_0 = 1$ has displacement rank 2 (except when $c_i = 0$ for $i = 1, \ldots, n-1$); it is easy to verify that

$$T - ZTZ^* = \begin{pmatrix} 1 & 0 \\ c_1 & c_1 \\ \vdots & \vdots \\ c_{n-1} & c_{n-1} \end{pmatrix} \begin{pmatrix} 1 & 0 \\ 0 & -1 \end{pmatrix} \begin{pmatrix} 1 & 0 \\ c_1 & c_1 \\ \vdots & \vdots \\ c_{n-1} & c_{n-1} \end{pmatrix}^*.$$

Hankel matrices also have displacement rank 2, with respect to the operator

$$\nabla H = H - ZHZ.$$

Example 2.2.2. Let T be an $n \times n$ real symmetric Toeplitz matrix, as in Example 2.2.1. Then the inverse of T also has displacement rank two. Indeed, a particular case of the formula of Gohberg and Semencul (see [61]) states that

$$
T^{-1} = \begin{pmatrix} a_0 & & \text{\Large O} \\ a_1 & a_0 & \\ \vdots & & \ddots \\ a_{n-1} & \cdots & \cdots & a_0 \end{pmatrix} \begin{pmatrix} a_0 & a_1 & \cdots & a_{n-1} \\ & a_0 & & \vdots \\ \text{\Large O} & & \ddots & \vdots \\ & & & a_0 \end{pmatrix}
$$

$$
- \begin{pmatrix} b_0 & & \text{\Large O} \\ b_1 & b_0 & \\ \vdots & & \ddots \\ b_{n-1} & \cdots & \cdots & b_0 \end{pmatrix} \begin{pmatrix} b_0 & b_1 & \cdots & b_{n-1} \\ & b_0 & & \vdots \\ \text{\Large O} & & \ddots & \vdots \\ & & & b_0 \end{pmatrix}
$$

for certain vectors $\mathbf{a} = [a_0, \ldots, a_{n-1}]^T$ and $\mathbf{b} = [b_0, \ldots, b_{n-1}]^T$. It follows that

$$
\begin{aligned}
T^{-1} - Z T^{-1} Z^T &= \mathbf{a}\mathbf{a}^T - \mathbf{b}\mathbf{b}^T \\
&= \begin{bmatrix} \mathbf{a} & \mathbf{b} \end{bmatrix} \begin{bmatrix} 1 & 0 \\ 0 & -1 \end{bmatrix} \begin{bmatrix} \mathbf{a}^T \\ \mathbf{b}^T \end{bmatrix}.
\end{aligned}
$$

The concept of displacement rank can be promptly extended to non-Hermitian matrices by defining ∇R in a suitable way and requiring that it should have low rank. Another definition of matrix displacement structure is found in [68], where the authors examine the *Sylvester-type operator*

$$
\nabla R = F R - R A^*, \tag{2.2.3}
$$

for suitable matrices F and A.

More generally, let Ω, Δ, F and A be $n \times n$ complex matrices and let $R \in \mathbb{C}^{n \times n}$ be such that

$$
\nabla_{\{\Omega, \Delta, F, A\}} R = \Omega R \Delta - F R A^* = G B^*, \tag{2.2.4}
$$

where $G, B \in \mathbb{C}^{n \times r}$. The pair $\{G, B\}$ is an $\{\Omega, \Delta, F, A\}$-*generator* of R and the smallest possible value for r among all the generators is called $\{\Omega, \Delta, F, A\}$-*displacement rank* of R. This definition generalizes both (2.2.1) and (2.2.3).

There clearly are many ways to define the matrices Ω, Δ, F and A, and therefore a displacement operator. The best choice for ∇R depends on the properties of R and on the type of problem that is to be solved.

We will be mainly interested in matrices which have displacement structure with respect to some particular cases of the operator defined in (2.2.3).

2.2.1. Toeplitz-like matrices

Let T be a Toeplitz matrix; it can be proved, as a generalization of Example 2.2.1, that T has displacement rank 2 with respect to the operator

$$\nabla R = Z_1 R - R Z_{-1}, \tag{2.2.5}$$

where

$$Z_\phi = \begin{pmatrix} 0 & \cdots & & \cdots & 0 & \phi \\ 1 & 0 & & \cdots & \cdots & 0 \\ 0 & 1 & \ddots & & & \vdots \\ \vdots & & \ddots & \ddots & & \vdots \\ 0 & \cdots & & 0 & 1 & 0 \end{pmatrix} \tag{2.2.6}$$

for $\phi - \pm 1$. Matrices which have low displacement rank with respect to (2.2.5) are called *Toeplitz like*.

2.2.2. Cauchy-like matrices

A *Cauchy-like matrix* is a matrix C with low displacement rank r with respect to the operator (2.2.3), where F and A are chosen as diagonal matrices. An $n \times n$ Cauchy-like matrix C has the form

$$C = \left[\frac{\mathbf{u}_i \mathbf{v}_j^*}{f_i - \bar{a}_j} \right]_{i,j=0}^{n-1}, \tag{2.2.7}$$

where $\mathbf{u}_i$ and $\mathbf{v}_j$ are row vectors of length r, whereas the f_i and a_j are complex scalars such that $f_i - \bar{a}_j \neq 0$ for all i, j. If F and A are chosen as

$$F = \begin{pmatrix} f_0 & & & \\ & f_1 & & \\ & & \ddots & \\ & & & f_{n-1} \end{pmatrix}, \quad A = \begin{pmatrix} a_0 & & & \\ & a_1 & & \\ & & \ddots & \\ & & & a_{n-1} \end{pmatrix},$$

then the displacement equation satisfied by C is

$$\nabla C = FC - CA^* = \begin{bmatrix} \mathbf{u}_0 \\ \mathbf{u}_1 \\ \vdots \\ \mathbf{u}_{n-1} \end{bmatrix} \begin{bmatrix} \mathbf{v}_0 \\ \mathbf{v}_1 \\ \vdots \\ \mathbf{v}_{n-1} \end{bmatrix}^*, \qquad (2.2.8)$$

i.e. C has displacement rank r with respect to the operator defined in (2.2.8).

Cauchy matrices correspond to the particular case of (2.2.7) when $r = 1$ and $\mathbf{u}_i = \mathbf{v}_i = 1$. Therefore Cauchy matrices have displacement rank 1 with respect to (2.2.8).

2.3. Computation of displacement generators

We collect here some results on the structure of sums, products and inverses of displacement structured matrices. These results prove useful when one needs to perform arithmetic operations on structured matrices, whose generators are known. Indeed, in such cases it is usually better to perform all operations on generators, rather than recovering the entries of the input matrices. See [19] for more details.

2.3.1. Sum of displacement structured matrices

Let ∇ be a displacement operator and let M_1 and M_2 be $n \times n$ matrices such that

$$\nabla M_1 = G_1 B_1, \qquad \text{with} \qquad G_1 \in \mathbb{C}^{n \times \alpha}, \ B_1 \in \mathbb{C}^{\alpha \times n}$$

and

$$\nabla M_2 = G_2 B_2, \qquad \text{with} \qquad G_2 \in \mathbb{C}^{n \times \beta}, \ B_2 \in \mathbb{C}^{\beta \times n}.$$

Then $M = M_1 + M_2$ is structured with respect to ∇, and its displacement rank is less than or equal to $\alpha + \beta$. Generators for M can be written as "union" of generators for M_1 and M_2; that is:

$$\nabla M = [G_1 \quad G_2] \cdot \begin{bmatrix} B_1 \\ B_2 \end{bmatrix}. \qquad (2.3.1)$$

These generators are clearly not minimal if the displacement rank of M is strictly less than $\alpha + \beta$.

2.3.2. Product of displacement structured matrices

The following general result is found in [19]; see also [102].

Proposition 2.3.1. *Let K, L, M and N be four fixed matrices and define the operators*

$$\nabla_1 A = A - KAL,$$
$$\nabla_2 B = B - MBN,$$
$$\nabla C = C - KCN.$$

Let $\Delta = LM - I$. Then the following equation holds:

$$\nabla(AB) = (\nabla_1 A)B + KAL(\nabla_2 B) + \Delta_0,$$

where $\Delta_0 = KA\Delta BN$.

2.3.3. Inverse of a displacement structured matrix

The invariance of displacement rank under inversion – anticipated in Example 2.2.2 – is one of the fundamental invariance properties of displacement structure theory. More details can be found for example in [19] and [79].

The inversion of displacement structured matrices is of course strictly related to the triangular factorization of such matrices and to the solution of linear systems, which may prove useful for the computation of polynomial GCDs. Special techniques for solving displacement structured systems rely on another fundamental invariance property: the Schur complements of a structured matrix inherit its displacement structure. Moreover, the generators of the Schur complement can be computed through a *generalized Schur algorithm*. This topic will be in part examined in the next section.

2.4. Fast GEPP for structured matrices

Let M be an $n \times n$ matrix having displacement rank α with respect to a fixed displacement operator. Then it follows from the definition of displacement structure that the n^2 entries of M are completely determined by $2\alpha n$ parameters - namely, the entries of the generators.

This fact suggests that it might be possible to design fast algorithms for computations involving structured matrices. This is actually the case, and in the literature several $\mathcal{O}(n^2)$-algorithms for the triangular factorization of structured matrices can be found. The first results in this direction are implicitly found in Schur's paper [119], and were greatly improved and generalized through the work of Kailath and colleagues (see [77, 78, 79]) and through the work of Heinig and Rost ([64]).

The main idea behind fast Gaussian elimination relies on the properties of Schur complements (which is the reason why implementations of fast Gaussian elimination for structured matrices are sometimes known as *generalized Schur algorithms*). Roughly speaking, Schur complements of a matrix with displacement structure still retain the same displacement structure. This allows to perform Gaussian elimination working on the generators, rather than on the whole matrix, thus reducing the number of operations needed.

Basic Gaussian elimination (*i.e.* without pivoting), however, is known to be quite unstable. Unfortunately, in most cases pivoting cannot be straightforwardly applied. Complete pivoting does not seem to be an option, because it involves n^3 comparisons and is therefore too expensive to be employed in a fast algorithm. Moreover, while it is true that Schur complementation preserves the displacement structure, this might not be the case for permutations. A permutation applied to a Toeplitz-like matrix, for example, generally disrupts the Toeplitz-like structure.

There is an exception, though, to the latter objection. It is easy to see that Cauchy-like matrices retain their form (2.2.7) after any permutation of columns and rows (of course corresponding permutations must be applied to the diagonal matrices that define the displacement operator in order to maintain the low displacement rank). So it is possible to incorporate pivoting techniques into fast algorithms for Cauchy-like matrices.

2.4.1. The GKO algorithm

The algorithm that we now present is due to Gohberg, Kailath and Olshevsky (see [60]), and is therefore known as *GKO algorithm*; it computes the Gaussian elimination with partial pivoting (GEPP) of a Cauchy-like matrix and can be extended to other classes of displacement structured matrices.

First of all, we need a lemma regarding the Schur complementation of a Cauchy-like matrix; indeed, recall that performing Gaussian elimination on an arbitrary matrix is equivalent to applying recursive Schur complementation.

Lemma 2.4.1. *Let the matrix*

$$C_1 = \begin{bmatrix} d_1 & u_1 \\ l_1 & C_{22} \end{bmatrix}$$

satisfy the Sylvester-type displacement equation

$$\nabla_{F_1, A_1}(C_1) = \begin{bmatrix} f_1 & 0 \\ * & F_2 \end{bmatrix} C_1 - C_1 \begin{bmatrix} a_1 & * \\ 0 & A_2 \end{bmatrix} = G_1 B_1, \qquad (2.4.1)$$

with $G_1 \in \mathbb{C}^{n \times \alpha}$, $B_1 \in \mathbb{C}^{\alpha \times n}$. If $d_1 \neq 0$, then the Schur complement

$$C_2 = C_{22} - \frac{1}{d_1} l_1 u_1$$

satisfies the Sylvester-type displacement equation

$$F_2 C_2 - C_2 A_2 = G_2 B_2,$$

with

$$\begin{bmatrix} 0 \\ G_2 \end{bmatrix} = G_1 - \begin{bmatrix} 1 \\ \dfrac{1}{d_1} l_1 \end{bmatrix} g_1, \qquad \begin{bmatrix} 0 & B_2 \end{bmatrix} = B_1 - b_1 \begin{bmatrix} 1 & \dfrac{1}{d_1} u_1 \end{bmatrix}, \quad (2.4.2)$$

where g_1 is the first row of G_1 and b_1 is the first column of B_1.

Proof. Recall the Schur complementation formula:

$$C_1 = \begin{bmatrix} 1 & 0 \\ \dfrac{1}{d_1} l_1 & I \end{bmatrix} \cdot \begin{bmatrix} d_1 & 0 \\ 0 & C_2 \end{bmatrix} \cdot \begin{bmatrix} 1 & \dfrac{1}{d_1} u_1 \\ 0 & I \end{bmatrix}. \qquad (2.4.3)$$

From (2.4.1) and (2.4.3) we have

$$\begin{bmatrix} f_1 & 0 \\ * & F_2 \end{bmatrix} \cdot \begin{bmatrix} d_1 & 0 \\ 0 & C_2 \end{bmatrix} - \begin{bmatrix} d_1 & 0 \\ * & C_2 \end{bmatrix} \cdot \begin{bmatrix} a_1 & * \\ 0 & A_2 \end{bmatrix}$$

$$= \begin{bmatrix} 1 & 0 \\ -\dfrac{1}{d_1} l_1 & I \end{bmatrix} \cdot B_1 \cdot G_1 \cdot \begin{bmatrix} 1 & -\dfrac{1}{d_1} u_1 \\ 0 & I \end{bmatrix}$$

and by equating the right lower blocks, the thesis follows.   □

Partial pivoting requires left multiplication by a permutation matrix before each elimination step, *i.e.*,

$$P_1 \cdot C_1 = \begin{bmatrix} 1 & 0 \\ \dfrac{1}{d_1} l_1 & I \end{bmatrix} \cdot \begin{bmatrix} d_1 & u_1 \\ 0 & C_2 \end{bmatrix}.$$

So, a step of the fast GEPP algorithm for a Cauchy-like matrix can be summarized as follows (we assume that the generators of the matrix are given):

- Use (2.2.7) to recover the first column $\begin{bmatrix} d_1 \\ l_1 \end{bmatrix}$ of C_1 from the generators.
- Determine the position (say, $(k, 1)$) of the entry of maximum magnitude in the first column.

- Let P_1 be the permutation matrix that interchanges the first and k-th rows. Interchange the first and k-th diagonal entries of F_1; interchange the first and k-th rows of G_1.
- Recover from the generators the first row $\begin{bmatrix} d_1 & u_1 \end{bmatrix}$ of $P_1 C_1$. Now one has the first column $\begin{bmatrix} 1 \\ \frac{1}{d_1} l_1 \end{bmatrix}$ of L and the first row $\begin{bmatrix} d_1 & u_1 \end{bmatrix}$ of U in the LU factorization of $P_1 C_1$.
- Use (2.4.2) to compute generators of the Schur complement C_2 of $P_1 \cdot C_1$.

Proceeding recursively, one obtains the factorization $C_1 = PLU$, where P is the product of the permutation matrices used in the process.

2.4.2. The Toeplitz-like case

Following an idea of Heinig (see [65]) and Pan (see [102]), this algorithm can be generalized to other structured matrices, such as Toeplitz-like, Toeplitz-plus-Hankel-like or Vandermonde-like (see again [60]). Indeed, such matrices can be transformed in a fast and stable way into Cauchy-like matrices. The following result concerns the Toeplitz-like case.

Proposition 2.4.2. *Let $R \in \mathbb{C}^{n \times n}$ be a Toeplitz-like matrix satisfying*

$$\nabla_{Z_1, Z_{-1}}(R) = Z_1 R - R Z_{-1} = GB,$$

where $G \in \mathbb{C}^{n \times \alpha}$, $B \in \mathbb{C}^{\alpha \times n}$ and Z_1, Z_{-1} are as in Section 2.2.1. Then

$$C = \mathcal{F} R D_0^{-1} \mathcal{F}^*$$

is a Cauchy-like matrix, i.e.,

$$\nabla_{D_1, D_{-1}}(C) = D_1 C - C D_{-1} = \hat{G} \hat{B}, \qquad (2.4.4)$$

where

$$\mathcal{F} = \frac{1}{\sqrt{n}} [e^{\frac{2\pi i}{n}(k-1)(j-1)}]_{k,j}$$

is the normalized $n \times n$ Discrete Fourier Transform matrix,

$$D_1 = \mathrm{diag}(1, e^{\frac{2\pi i}{n}}, \ldots, e^{\frac{2\pi i}{n}(n-1)}),$$
$$D_{-1} = \mathrm{diag}(e^{\frac{\pi i}{n}}, e^{\frac{3\pi i}{n}}, \ldots, e^{\frac{(2n-1)\pi i}{n}}),$$
$$D_0 = \mathrm{diag}(1, e^{\frac{\pi i}{n}}, \ldots, e^{\frac{(n-1)\pi i}{n}}),$$

and

$$\hat{G} = \mathcal{F}G, \qquad \hat{B}^* = \mathcal{F} D_0 B^*.$$

Proof. Consider the following well-known factorizations (see Section B.3):

$$Z_1 = \mathcal{F}^* D_1 \mathcal{F}, \qquad Z_{-1} = D_0^{-1} \mathcal{F}^* D_{-1} D_0. \qquad (2.4.5)$$

Substituting (2.4.5) into (2.4.4) and multiplying by $\mathcal{F}$ on the left and by $D_0^{-1} \mathcal{F}^*$ on the right, one obtains the assertion of the proposition. □

Proposition 2.4.2 suggests an $\mathcal{O}(n^2)$ algorithm for solving linear systems with Toeplitz-like matrices. Suppose that the linear system

$$R\mathbf{x} = \mathbf{v}$$

is to be solved, where R is Toeplitz-like. Using Proposition 2.4.2, compute generators for the associated Cauchy-like matrix C. The whole matrix is not required; it is only necessary to apply FFT twice to the generators of R (see Section B.2 on the relationship between FFT and Fourier matrices). Then the fast GEPP algorithm provides a permuted LU factorization for C, so that we have

$$R = \mathcal{F}^* P L U \mathcal{F} D_0$$

and the linear system with matrix R is solved using forward and back substitution, plus two FFTs.

2.5. The Sylvester matrix

Let polynomials $u(x) = \sum_{i=0}^{n} u_i x^i$ and $v(x) = \sum_{i=0}^{m} v_i x^i$, of degrees n and m, respectively, be given. The Sylvester matrix of $u(x)$ and $v(x)$ is the $(m+n) \times (m+n)$ matrix

$$S(u,v) = \begin{pmatrix} u_n & u_{n-1} & \cdots & \cdots & u_0 & & & O \\ & \ddots & \ddots & & & & \ddots & \\ O & & u_n & u_{n-1} & \cdots & \cdots & u_0 & \\ v_m & v_{m-1} & \cdots & v_0 & & & & O \\ & \ddots & \ddots & & & \ddots & & \\ & & \ddots & \ddots & & & \ddots & \\ O & & & v_m & v_{m-1} & \cdots & v_0 \end{pmatrix}. \qquad (2.5.1)$$

Notice that some authors define the Sylvester matrix as the transpose of (2.5.1).

Remark 2.5.1. The Sylvester matrix has a Toeplitz-block structure. Moreover, it is displacement structured and has displacement rank at most 2 with respect to the operator (2.2.5).

The rank of a Sylvester matrix gives information on the degree of the GCD of the associated polynomials:

Theorem 2.5.2. (see e.g. [88]) *Let $g(x)$ be the GCD of $u(x)$ and $v(x)$. Then:*

- *$S(u, v)$ is nonsingular if and only if $u(x)$ and $v(x)$ are coprime;*
- *if $S(u, v)$ is singular, then $\dim \ker S(u, v) = \deg g$.*

The following theorem (see [88, 41, 121, 58]) shows how the coefficients of the GCD can be retrieved from the Sylvester matrix. The proof is useful to motivate the discussion on the approximate case in Section 4.3, because it describes the structure of the null space of $S(u, v)$.

Theorem 2.5.3. *Let $S(u, v) = QR$ be the QR factorization of the Sylvester matrix associated with polynomials $u(x)$ and $v(x)$. Then the last nonzero row of R gives the coefficients of a (possibly non monic) GCD of $u(x)$ and $v(x)$.*

Proof. From the definition of the Sylvester matrix it follows that

$$\begin{bmatrix} x^{m-1}u(x) \\ \vdots \\ u(x) \\ x^{n-1}v(x) \\ \vdots \\ v(x) \end{bmatrix} = Q \cdot R \cdot \begin{bmatrix} x^{n+m-1} \\ \vdots \\ x^{n-1} \\ \vdots \\ x \\ 1 \end{bmatrix} = Q \cdot \begin{bmatrix} r_{n+m-1}(x) \\ \vdots \\ r_d(x) \\ 0 \\ \vdots \\ 0 \end{bmatrix},$$

where each $r_i(x)$ is a polynomial of degree i whose coefficients are given by the entries in the $(n + m - i)$-th nonzero row of R for $i = d, d + 1, \ldots, n + m - 1$.

Suppose that x_k is a common root of $u(x)$ and $v(x)$, with multiplicity e_k. Define the $(n+m-1) \times e_k$ matrix Λ, parametrized by x_k, as follows:

$$\Lambda = \begin{bmatrix} x_k^{n+m-1} & (n+m-1)x_k^{n+m-2} & \cdots & (n+m-1)_{(e_k-1)}x_k^{n+m-e_k} \\ x_k^{n+m-2} & (n+m-2)x_k^{n+m-3} & \cdots & \vdots \\ \vdots & \vdots & & \\ \vdots & \vdots & & (e_k-1)! \\ x_k^2 & 2x_k & & \\ x_k & 1 & & \\ 1 & & & \end{bmatrix}$$

where the Pochhammer symbol for the falling factorial has been used for the entry in the right upper corner, i.e., $(n + m - 1)_{(e_k-1)} = (n + m - 1) \cdot \ldots \cdot (n + m - e_k + 1)$. Observe that, for $1 < j \le e_k$, the entries of the j-th column in the matrix Λ are the derivatives of the corresponding entries of the $(j - 1)$-th column with respect to x_k.

Then it is easy to see that $S\Lambda$ is the zero matrix of size $(n + m - 1) \times e_k$. Therefore $r_d(x)$ and all its derivatives up to order $e_k - 1$ are zero at x_k.

Conversely, assume that $r_d(x)$ and all its derivatives up to order $e_k - 1$ are zero at x_k. Then it follows from the upper triangular structure of R that $R\Lambda$ is zero.   □

Remark 2.5.4. As an alternative argument for Theorem 2.5.3, one may also observe that, since $R = Q^* S(u, v)$, then the rows of the triangular factor R are obtained as linear combinations of the rows of $S(u, v)$. This is equivalent to saying that the polynomials $r_i(x)$ defined above are linear combinations of the polynomials $x^{m-1}u(x), \ldots, u(x), x^{n-1}v(x), \ldots, v(x)$, and $r_d(x)$ is a polynomial of minimum degree that can be expressed in such a way; therefore $r_d(x)$ is a GCD of $u(x)$ and $v(x)$.

Notice that the QR factorization of $S(u, v)$ in Theorem 2.5.3 might not be unique if $u(x)$ and $v(x)$ are not coprime - that is, if $S(u, v)$ is singular. However, the proof shows that Theorem 2.5.3 holds for any QR factorization of $S(u, v)$. In other words, the possible nonuniqueness of R is limited to the first $n + m - d - 1$ rows, whereas it is guaranteed that the last d rows of R are zero. Indeed, a nonzero row among the last d rows of R would provide a polynomial of degree less than d, which can be written as a linear combination of $u(x)$ and $v(x)$. This is not possible, since it is assumed that $u(x)$ and $v(x)$ have a GCD of degree d.

The proof of Theorem 2.5.3 suggests a parametrization of the null space of the Sylvester matrix using the common roots of $u(x)$ and $v(x)$:

Proposition 2.5.5. Let $\{x_i\}_{i=1,\ldots,k}$ be the common roots of $u(x)$ and $v(x)$ and let i_m be the multiplicity of each common root. For every $i = 1, \ldots, k$ define i_m vectors of length $n + m$ as follows:

$$\mathbf{w}_1^{(i)} = [1, x_i, x_i^2, \ldots x_i^{n+m-1}]^T,$$

$$\mathbf{w}_2^{(i)} = [0, 1, 2x_i, \ldots (n + m - 1)x_i^{n+m-2}]^T,$$

$$\ldots$$

$$\mathbf{w}_{i_m}^{(i)} = [0, \ldots 0, i_m x_i, \ldots, (n + m - 1) \cdots (n + m - i_m + 1)x_i^{n+m-i_m}]^T.$$

Then the set of all vectors thus defined is a set of generators for the null space of $S(u, v)$.

Information on the GCD degree can also be obtained from *Sylvester subresultants*, which are submatrices of the (transpose) Sylvester matrix and are defined as

$$
S_j(u, v) =
\begin{pmatrix}
u_0 & & & v_0 & & \\
u_1 & \ddots & u_0 & v_1 & \ddots & v_0 \\
\vdots & \ddots & u_1 & \vdots & \ddots & v_1 \\
\vdots & & \vdots & \vdots & & \vdots \\
u_n & & \vdots & v_m & & \vdots \\
& \ddots & \vdots & & \ddots & \vdots \\
& & u_n & & & v_m \\
& \underbrace{}_{j} & & \underbrace{}_{j} &
\end{pmatrix}
\tag{2.5.2}
$$

for $j = 0, 1, \ldots, n - 1$. Under the hypotheses of Theorem 2.5.2, we find that:

- $S_j(u, v)$ has full (column) rank for $j < n - k$;
- the null space of $S_{n-k}(u, v)$ has dimension 1;
- if $z \in \ker S_{n-k}(u, v)$, then $z = [q_0 \cdots q_{m-k} \; p_0 \cdots p_{n-k}]^T$, where $p(x) = u(x)/g(x)$ and $q(x) = v(x)/g(x)$.

Lastly, we give an estimate on the norm of the Sylvester matrix.

Lemma 2.5.6. *Let $\|S(u,v)\|_F$ denote the Frobenius norm of the Sylvester matrix associated with $u(x)$ and $v(x)$. Then*

$$
\|S(u, v)\|_2 \le \|S(u, v)\|_F \le \sqrt{m\|u(x)\|_2^2 + n\|v(x)\|_2^2}.
$$

The proof easily follows from standard matrix norm inequalities (see Section A.2) and from the structure of the Sylvester matrix.

2.6. The Bézout matrix

Let $u(x) = \sum_{i=0}^n u_i x^i$ and $v(x) = \sum_{i=0}^m v_i x^i$ be univariate polynomials of degrees n and m respectively. Let

$$
b(x, y) = \frac{u(x)v(y) - u(y)v(x)}{x - y}.
$$

The rational function $b(x, y)$ is easily proved to be a polynomial in two variables and can therefore be written as

$$b(x, y) = \sum_{i,j} b_{ij} x^i y^j.$$

The matrix

$$B(u, v) = (b_{ij})_{ij}$$

is called the *Bézout matrix* or *Bezoutian* associated with $u(x)$ and $v(x)$.

Though arising from the theory of resultants, Bezoutians play an important role in many fields of numerical and symbolic computing, including signal processing and control theory (see e.g. [6, 69, 55]).

The sources for this presentation are mostly [19, 69, 89]. See also [66] for a review of classical and generalized forms of Bezoutians.

Let us now recall some essential properties of the Bézout matrix:

Theorem 2.6.1. *The following properties hold:*

1. $B(u, v)$ *is an $N \times N$ symmetric matrix, where $N = \max\{n, m\}$;*
2. $B(u, v) = -B(v, u)$;
3. $B(u, v)$ *is linear in u and v, that is*

$$B(\alpha u + \beta f, v) = \alpha B(u, v) + \beta B(f, v),$$
$$B(u, \alpha v + \beta g) = \alpha B(u, v) + \beta B(v, g),$$

for any polynomials $f(x)$, $g(x)$ and scalars α, β.

For ease of notation, we will assume from now on that $n \geq m$, that is, $N = n$.

Remark 2.6.2. Let J be as in (2.1.1) and let $\tilde{u}(x) = \sum_{i=0}^{n} u_{n-i} x^i$ and $\tilde{v}(x) = \sum_{i=0}^{m} v_{m-i} x^i$ be the reversed polynomials. Then

$$B(\tilde{u}, \tilde{v}) = J B(u, v) J.$$

We next state more different representations of the Bézout matrix.

Proposition 2.6.3. (matrix representation) *The Bezoutian matrix can be written as:*

$$
B(u, v) =
\begin{pmatrix}
u_1 & \cdots & u_n \\
\vdots & \ddots & \\
u_n & & 0
\end{pmatrix}
\begin{pmatrix}
v_0 & \cdots & v_{n-1} \\
& \ddots & \vdots \\
0 & & v_0
\end{pmatrix}
$$
$$
-
\begin{pmatrix}
v_1 & \cdots & v_n \\
\vdots & \ddots & \\
v_n & & 0
\end{pmatrix}
\begin{pmatrix}
u_0 & \cdots & u_{n-1} \\
& \ddots & \vdots \\
0 & & u_0
\end{pmatrix}
$$

where if $m < n$ the coefficients $v_{m+1}, \ldots, v_n$ are assumed to be zero.

As a consequence of the previous result, we have:

Proposition 2.6.4. (recursive formula) *The entries of $B(u, v)$ can be recursively calculated as follows:*

$$b_{i,j+1} = b_{i+1,j} + u_i v_j - u_j v_i,$$

setting $b_{n+1,j} = b_{0,j} = 0$ for all i and j and assuming $v_{m+1} = \cdots = v_n = 0$ as above.

The results listed so far show how to compute $B(u, v)$ given the coefficients of $u(x)$ and $v(x)$. One might also want to solve the inverse problem: given a Bezoutian B, find polynomials $u(x)$ and $v(x)$ such that $B = B(u, v)$. The solution of the corresponding problem for the Sylvester matrix is straightforward; for the Bezoutian, it is slightly more complicated.

Remark 2.6.5. Given the m-th and the last row of $B(u, v)$, together with the top left entry, it is possible to compute the coefficients of the polynomials $u_n v(x)$ and $u(x)/u_n$.

Indeed, if $m < n$, the last row of $B(u, v)$ is $u_n[v_0, \ldots, v_{n-1}]$, which define the coefficients of $u_n v(x)$. Moreover, the m-th row $\mathbf{b}^T$ of $B(u, v)$ is

$$\mathbf{b}^T = -v_m[u_0, \ldots, u_{n-1}] + [u_m, \ldots, u_n, 0, \ldots, 0] \begin{pmatrix} v_0 & \cdots & v_{n-1} \\ & \ddots & \vdots \\ & & v_0 \end{pmatrix}$$

and this relation can be used to express the coefficients of $u(x)/u_n$ as linear functions of u_{n-1}. The coefficient u_{n-1} may then be retrieved from the top left entry of $B(u, v)$. Observe that the hypothesis $m < n$ does not cause any loss of generality, because, if $m = n$, then $B(u, v) = B(u \bmod v)$.

We compute here an estimate on the norm of the Bezoutian that will later prove useful:

Lemma 2.6.6. *Let $u(x)$ and $v(x)$ be polynomials of degree n and m respectively, with $n \geq m$. Then*

$$\|\mathrm{Bez}(u, v)\|_2 \leq 2n \|u\|_2 \|v\|_2. \tag{2.6.1}$$

Proof. A bound for each entry of the Bezoutian can be obtained from the matrix representation:

$$|B(u, v)_{i,j}| = |B(u, v)_{(i,:)}[\underbrace{0, \dots, 0, 1, 0 \dots, 0}_{j-1}]^T| \leq 2\|u\|_2\|v\|_2. \quad (2.6.2)$$

Standard norm inequalities yield:

$$\|B(u, v)\|_2 \leq \|B(u, v)\|_F \leq 2\sqrt{n}\sqrt{n}\|u\|_2\|v\|_2, \quad (2.6.3)$$

and hence the thesis. $\qquad\square$

We are now going to state a number of results which show how Bezoutians are related to the polynomial GCD. The first two theorems can be found in [69].

Theorem 2.6.7. *Given polynomials $u(x)$ and $v(x)$, the following properties hold:*

- *$B(u, v)$ is invertible if and only if $u(x)$ and $v(x)$ are coprime;*
- *moreover, $\dim(\ker B(u, v))$ is equal to the degree of $GCD(u, v)$.*

Theorem 2.6.8. *Given a polynomial $w(x) = \sum_{i=0}^{k} w_i x^i$, the following holds:*

$$B(uw, vw) = C(w)B(u, v)C(w)^T,$$

where:

$$C(w) = \begin{pmatrix} w_0 & & & \\ w_1 & \ddots & & \\ \vdots & \ddots & w_0 & \\ \vdots & & w_1 & \\ w_k & & \vdots & \\ & \ddots & \vdots & \\ & & w_k & \end{pmatrix}$$

is a convolution matrix of appropriate size.

Remark 2.6.9. As a particular case of Theorem 2.6.8, let $g(x) = \sum_{i=0}^{k}$ be the greatest common divisor of $u(x)$ and $v(x)$, and let $\hat{u}(x)$ and $\hat{v}(x)$ be such that

$$u(x) = \hat{u}(x)g(x),$$
$$v(x) = \hat{v}(x)g(x).$$

Then we have

$$B(u, v) = GB(\hat{u}, \hat{v})G^T, \quad (2.6.4)$$

where

$$G = \begin{pmatrix} g_0 & & & & \\ g_1 & \ddots & & & \\ \vdots & & \ddots & & g_0 \\ \vdots & & & & g_1 \\ g_k & & & & \vdots \\ & & \ddots & & \vdots \\ & & & & g_k \end{pmatrix}$$

is a $n \times (n - k)$ matrix.

The null space of $B(u, v)$ can be characterized exactly like in Proposition 2.5.5. As particular case, we have:

Remark 2.6.10. Let $\alpha_1, \ldots \alpha_k$ be the common roots of u and v and suppose that such roots are distinct. Then a basis for the null space of $B(u, v)$, which has dimension k, is given by $\{[1, \alpha_i, \alpha_i^2, \ldots, \alpha_i^{n-1}]^T\}_{i=1,\ldots k}$.

We prove here that a result similar to Theorem 2.5.3 also holds for the Bezoutian:

Theorem 2.6.11. *Let B be the Bézout matrix associated with the polynomials $u(x)$ and $v(x)$ and let J be as in (2.1.1). If $JBJ = QR$ is the QR decomposition of the permuted Bezoutian, then the entries in the last row of R are the coefficients of a GCD of $u(x)$ and $v(x)$.*

Proof. The theorem can be proved using an argument similar to Remark 2.5.4. Recall from Remark 2.6.2 that $JBJ = B(\tilde{u}, \tilde{v})$, where $\tilde{u}(x)$ and $\tilde{v}(x)$ are the reversed polynomials. Observe that $R = Q^* B(\tilde{u}, \tilde{v})$. Then it follows from the factorization (2.6.4) applied to $B(\tilde{u}, \tilde{v})$ that the rows of R are obtained as linear combinations of the rows of G, which contain the coefficients of a GCD $g(x)$. In terms of polynomials, this means that, if $r_i(x), i = d, \ldots n-1$ is the polynomial of degree i whose coefficients are given by the $(n - i)$-th row of R, then the $r_i(x)$'s are linear combinations of $x^{n-k-1}g(x), \ldots, xg(x), g(x)$ and $r_d(x)$ is a polynomial of minimum degree among such linear combinations. Therefore $r_d(x)$ must be a scalar multiple of $g(x)$. $\qquad\square$

The remarks on the possible nonuniqueness of the QR decomposition of the Sylvester matrix apply also to the Bezoutian. The QR decomposition of $B(u, v)$ might not be unique if $B(u, v)$ is singular; however, any QR decomposition provides a GCD as described in Theorem 2.6.11.

The following two theorems are particular cases of Theorems 6.2 and 6.3 of [3] and allow to find submatrices of the Bézoutian which behave in a similar way to the subresultants defined for Sylvester matrices (see [47])

Theorem 2.6.12. *If* $c_1, \ldots, c_n$ *are the columns of the matrix* $B(u, v)$ *and its rank is* $n-k$, *then the last* $n-k$ *columns* $c_{k+1}, \ldots, c_n$ *are linearly independent and each* c_i $(1 \leq i \leq k)$ *can be written as a linear combination of* $c_{k+1}, \ldots, c_n$.

Theorem 2.6.13. *Let* $c_1, \ldots, c_n$ *be the columns of the matrix* $B(u, v)$ *and* $n - k$ *its rank. Write each* c_i $(1 \leq i \leq k)$ *as a linear combination of* $c_{k+1}, \ldots, c_n$ *as follows:*

$$c_{k-i} = h_{k-i}^{k+1} c_{k+1} + \sum_{j=k+2}^{n} h_{k-i}^j c_j, \qquad i = 0, \ldots, k-1.$$

Let $d_1, \ldots, d_k$ *be given by*

$$d_j = d_0 h_{k-j+1}^{k+1},$$

where $d_0 \in F$. *Then*

$$D(x) = d_0 x^k + d_1 x^{k-1} + \cdots + d_{k-1} x + d_k$$

is a GCD for $u(x)$ *and* $v(x)$.

2.7. More results on resultant matrices

Let F_u be the Frobenius (or companion) matrix associated with the monic polynomial $u(x)$:

$$F_u = \begin{pmatrix} 0 & 1 & 0 & \cdots & 0 \\ 0 & 0 & 1 & \ddots & \vdots \\ \vdots & & \ddots & \ddots & 0 \\ 0 & \cdots & \cdots & 0 & 1 \\ -u_0 & -u_1 & \cdots & \cdots & u_{n-1} \end{pmatrix}.$$

The following result, known as *Barnett factorization* ([6]), establishes a relationship between the Bézoutian of $u(x)$ and $v(x)$ and the matrix polynomial $v(F_u)$.

Theorem 2.7.1. *The Bézout matrix associated with* $u(x)$ *and* $v(x)$ *can be factorized as*

$$B(u, v) = B(u, 1)v(F_u),$$

where $v(F_u) = \sum_{i=0}^{m} v_i (F_u)^m$ *and* $u(x)$ *is assumed to be monic.*

Lemma 2.7.2. *Let $u(x)$ and $v(x)$ be two monic polynomials of degrees n, m respectively, with $m < n$. Then $v(F_u^T)\mathbf{z} = \mathbf{0}$ if and only if $v(x)z(x) = 0 \mod u(x)$, where $\mathbf{z}$ is the vector of coefficients of $z(x)$. Therefore, $v(F_u^T)$ is singular if and only if $g(x) = \mathrm{GCD}(u(x), v(x))$ is a nonconstant polynomial.*

Moreover, if $u(x) = w(x)g(x)$, the null space of $v(F_u^T)$ is

$$\ker v(F_u^T) = \{\mathbf{z} \in \mathbb{C}^n : z(x) = p(x)w(x) \text{ for some } p(x)\}$$

and it has dimension $\deg g(x)$.

Proof. Denote with $\mathbf{e}^{(i)}$, for $i = 0, \ldots, n - 1$ the $(i + 1)$-th vector of the canonical basis in $\mathbb{C}^n$. It is easy to see that $(F_u^T)^i \mathbf{e}^{(0)}$ for all i, so that the first column of $v(F_u^T)$ is $[v_0, v_1, \ldots, v_m\, 0, \ldots, 0]^T$. The same observation applied to polynomials $z(x)$ and $r(x)$, where $r(x) = v(x)z(x) \mod u(x)$ has degree less than n, shows that

$$v(F_u^T)\mathbf{z} = v(F_u^T)z(F_u^T)\mathbf{e}^{(0)} = r(F_u^T)\mathbf{e}^{(0)}.$$

It follows that $v(F_u^T)\mathbf{z} = \mathbf{0}$ if and only if $v(x)z(x) = 0 \mod u(x)$. In particular, $z(x)$ must be a multiple of $w(x)$. □

Theorem 2.7.1 and Lemma 2.7.2 show that $v(F_u)$ may be regarded as a kind of resultant matrices - it is sometimes called *companion matrix resultant* in the literature - and used in the study of polynomial GCD. Also notice that $v(F_u)$ is displacement structured.

We are now going to introduce another matrix associated with a given pair of polynomials, which also shares some properties of resultant matrices. Assume that $u(x)$ and $v(x)$ are given as above, with $m < n$. Consider the reverse polynomials $\hat{u}(x) = x^n u(x^{-1})$ and $\hat{v}(x) = x^m v(x^{-1})$ and define the rational function

$$h(x) = x^{n-m-1} \frac{\hat{v}(x)}{\hat{u}(x)} = \frac{v(x^{-1})}{x u(x^{-1})}$$

and write $h(x)$ as a formal power series

$$h(x) = \sum_{i=0}^{+\infty} h_i x^i. \qquad (2.7.1)$$

The coefficients of (2.7.1) are related to the coefficients of $u(x)$ and $v(x)$ through the following triangular Toeplitz linear system:

$$
\begin{pmatrix}
u_n & & & \\
u_{n-1} & u_n & & \\
u_{n-2} & u_{n-1} & u_n & \\
\ddots & \ddots & \ddots & \ddots
\end{pmatrix}
\begin{pmatrix}
h_0 \\
h_1 \\
h_2 \\
\vdots
\end{pmatrix}
=
\begin{pmatrix}
v_{n-1} \\
v_{n-2} \\
v_{n-3} \\
\vdots
\end{pmatrix},
\qquad (2.7.2)
$$

with the convention that $u_j = v_j = 0$ for $j < 0$ and $v_i = 0$ for $i > m$.

Observe that the coefficients of (2.7.1) (which are also known as *Markov parameters*) can be used to write the power series expansion of the function $R(x) = v(x)/u(x)$ as well:

$$
R(x) = \frac{v(x)}{u(x)} = \sum_{i=0}^{+\infty} h_i x^{-i-1}. \qquad (2.7.3)
$$

We can now define the $n \times n$ Hankel matrix $H(u, v)$ as follows:

$$
H(u, v)_{i,j} = h_{i+j}.
$$

If the coefficients of $u(x)$ and $v(x)$ are given, then the entries of $H(u, v)$ can be computed by solving the linear system given by the first $2n - 1$ equations of (2.7.2).

Proposition 2.7.3. *The Bézout and Hankel matrices associated with a pair of polynomials $u(x)$ and $v(x)$ are related through the following result (see [19, 68]):*

$$
B(u, v) = B(u, 1)H(u, v)B(u, 1).
$$

Remark 2.7.4. It follows from Theorem 2.7.1 and Proposition 2.7.3 that

$$
H(u, v) = v(F_u)B(u, 1)^{-1}.
$$

The next proposition gives further details on the relationship between the GCD of a pair of polynomials and the associated Hankel matrix; also observe that the first statement is closely related to Kronecker's theorem for Hankel operators. The $j \times j$ leading principal submatrix of a matrix A will be denoted as A_j in the following.

Proposition 2.7.5. *Let $u(x)$ and $v(x)$ be two monic polynomials of degrees n, m respectively, with $m < n$, and let $g(x) = \mathrm{GCD}(u(x), v(x))$ be a monic polynomial of degree $n - k$. Then:*

- $\operatorname{rank} H(u, v) = k$, $\det H_k \neq 0$, *whereas* $\det H_i = 0$ *for* $i > k$;
- *if* $H_{k+1}\mathbf{w} = \mathbf{0}$ *for* $\mathbf{w} = [w_0 \ \ldots \ w_{k-1} \ 1]^T$, *that is, if*

$$
H_k \begin{pmatrix} w_0 \\ \vdots \\ w_{k-1} \end{pmatrix} = - \begin{pmatrix} h_k \\ \vdots \\ h_{2k-1} \end{pmatrix},
$$

then

$$
u(x) = g(x) \sum_{i=0}^{k} w_i x^i.
$$

Proof. Assume that $u(x) = w(x)g(x)$ and $v(x) = t(x)g(x)$, where $w(x)$ and $t(x)$ are relatively prime. It follows from Remark 2.7.4 that $H(u, v)\mathbf{z} = \mathbf{0}$ if and only if $v(F_u)^T\mathbf{z} = \mathbf{0}$. Now, Lemma 2.7.2 says that $v(F_u)^T\mathbf{z} = \mathbf{0}$ holds if and only if $z(x)$ can be factorized as $z(x) = p(x)w(x)$ for some polynomial $p(x)$, that is, if $z(x)$ belongs to

$$
\mathcal{K} = \{a(x) \in \mathbb{C}(x) : \deg a(x) = n, \ a(x) = p(x)w(x), \deg p(x) \leq n - k - 1\}.
$$

But $\mathcal{K}$ is a linear space of dimension $n - k$ and therefore $\operatorname{rank} H(u, v) = \operatorname{rank} v(F_u^T) = k$. This proves the first assertion. The second assertion is satisfied by choosing $p(x) = 1$. $\qquad\square$

The above proposition can be rewritten in terms of the Bezoutian matrix as follows (see [15]).

Corollary 2.7.6. *In the hypotheses of Proposition 2.7.5, we have:*

- *rank* $B(u, v) = k$, $\det(J B(u, v)J)_k \neq 0$, *whereas* $\det(J B(u, v)J)_i = 0$ *for* $i > k$;
- *if* $(J B(u, v)J)_{k+1}\mathbf{y} = \mathbf{0}$ *for* $\mathbf{y} = [y_0 \ \ldots \ y_{k-1} \ y_k]^T$, *then*

$$
u(x) = g(x) \sum_{i=0}^{k} w_i x^i,
$$

where $\mathbf{w} = (B(u, 1)J)_{k+1}\mathbf{y} = [w_0 \ \ldots \ w_k]^T$.

The previous results allow to formulate the following algorithm for polynomial GCD:

Algorithm BézoutGCD (Algorithm 9.1 in [19])

Input: the coefficients of polynomials $u(x) = \sum_{i=0}^{n} u_i x^i$ and $v(x) = \sum_{i=0}^{m} v_i x^i$.
Output: the coefficients of $s(x) = \sum_{i=0}^{k} s_i x^i = \mathrm{GCD}(u, v)$.

1. Compute $\operatorname{rank} B(u, v)$ as follows:

   (i) compute the coefficients $\{c_i\}_{i=0,1,\dots,n-1}$ of the characteristic polynomial of $B(u, v)$, by exploiting the displacement structure of the Bezoutian matrix;

   (ii) set k such that $c_{n-k} \neq 0$ and $c_i = 0$, $i = 0, \dots, n - k - 1$.

2. Solve the $k \times k$ system $C\mathbf{y} = \mathbf{b}$, where $C = (JB(u, v)J)_k$ and

$$
\mathbf{b} = \begin{pmatrix} v_n & & 0 \\ \vdots & \ddots & \\ v_{n-k+1} & \cdots & v_n \end{pmatrix} \begin{pmatrix} u_{n-k+1} \\ \vdots \\ u_{n-2k} \end{pmatrix}
$$

$$
- \begin{pmatrix} u_n & & 0 \\ \vdots & \ddots & \\ u_{n-k+1} & \cdots & u_n \end{pmatrix} \begin{pmatrix} v_{n-k+1} \\ \vdots \\ v_{n-2k} \end{pmatrix}.
$$

3. Compute $\mathbf{w} = [w_1, \dots, w_k]^T = (B(u, 1)J)_{k+1} \begin{pmatrix} y_0 \\ \vdots \\ y_{k-1} \\ 1 \end{pmatrix}$.

4. Compute $s(x)$ such that $u(x) = s(x)w(x)$.

Remark 2.7.7. The above algorithm could be reformulated in terms of the Hankel matrix $H(u, v)$ associated with the input polynomials. However, the Bezoutian matrix allows better numerical stability, because its entries are bounded in modulus by $\phi = 2n\mu v$ (where $\|u(x)\|_\infty \leq \mu$ and $\|v(x)\|_\infty \leq v$), whereas the entries of $H(u, v)$ may grow exponentially with n.

Moreover, if $u(x)$ and $v(x)$ have integer coefficients, the whole computation can still be kept within the integers. See [19] for details.

Chapter 3
The Euclidean algorithm

The Euclidean algorithm is probably the oldest and most widely known method for GCD computations and there are a number of efficient Euclidean-based methods for computing polynomial GCDs, beginning with the work of Collins and Brown (see [38, 25, 26]).

Several papers have been devoted to the study and adaptation of some of the variants of the Euclidean algorithm to the approximate GCD problem (see [118, 100, 10, 11, 74, 115]). Most versions have a computational cost that is quadratic in the degree of the polynomials and are therefore quite cheap if compared to other approches to approximate GCD computations.

Stability issues, however, must be taken into account and carefully analyzed, because the classical algorithm turns out to be unstable when applied to approximate polynomials. Stabilized versions of the Euclidean algorithm are mainly based either on modified termination criteria, or on clever look-ahead techniques that allow to "jump" over ill-conditioned subproblems.

Moreover, it should be pointed out that any version of the Euclidean algorithm can only be guaranteed to yield a lower bound on the approximate GCD degree and may fail to give the "correct" degree. This is because the Euclidean algorithm is basically equivalent to performing Gaussian elimination on the Sylvester matrix with a fixed elimination sequence and therefore poses excessive constraints on the perturbations that are allowed to enter the computation (see [121] for a discussion).

3.1. The classical algorithm

Recall that the classical algorithm for the computation of polynomial GCD goes as follows. Let polynomials $u(x)$ and $v(x)$ be given, with $\deg u(x) \geq \deg v(x)$; then there exist polynomials $q_1(x)$ (*quotient*) and $r_1(x)$ (*remainder*) such that

$$u(x) = q_1(x)v(x) + r_1(x)$$

and $\deg r_1(x) < \deg v(x)$. If $r_1(x) = 0$, then $v(x) = \text{GCD}(u(x), v(x))$. Otherwise, observe that $\text{GCD}(u(x), v(x)) = \text{GCD}(v(x), r_1(x))$ and perform a second polynomial division:

$$v(x) = q_2(x)r_1(x) + r_2(x).$$

Again, if $r_2(x) = 0$, then $r_1(x) = \text{GCD}(u(x), v(x))$; otherwise compute

$$r_1(x) = q_3(x)r_2(x) + r_3(x)$$

and so on. In other words, the Euclidean algorithm yields a sequence of polynomials $\{r_i(x)\}_{i=-1,0,1,\dots,s}$, where

$$
\begin{aligned}
r_1(x) &= u(x), \\
r_0(x) &= v(x), \\
r_i(x) &= q_{i+1}(x)r_{i+1}(x) + r_{i+2}(x)
\end{aligned}
\tag{3.1.1}
$$

for some polynomials $q_i(x)$, and $\deg r_{i+1}(x) < \deg r_i(x)$. The algorithm stops when a remainder $r_s(x) = 0$ is found; then $r_{s-1}(x)$ is the desired GCD of $u(x)$ and $v(x)$. Observe that, since the degrees of the $r_i(x)$'s are strictly decreasing, the algorithm must terminate in a finite number of steps.

3.1.1. Matrix representation

If the recursive equations (3.1.1) are written in the form of vector and matrix equations, we obtain the *matrix representation* of the Euclidean algorithm:

$$
\begin{bmatrix} r_{i+1}(x) \\ r_i(x) \end{bmatrix} = \tilde{Q}_i(x) \begin{bmatrix} r_i(x) \\ r_{i-1}(x) \end{bmatrix} = \hat{Q}_i(x) \begin{bmatrix} u(x) \\ v(x) \end{bmatrix},
\tag{3.1.2}
$$

$$
\tilde{Q}_i(x) = \begin{bmatrix} 0 & 1 \\ 1 & -q_i(x) \end{bmatrix},
\tag{3.1.3}
$$

$$
\hat{Q}_i(x) = \tilde{Q}_i(x)\tilde{Q}_{i-1}(x)\dots\tilde{Q}_1(x), \quad i = 1, \dots, l, \quad l \le n+1.
\tag{3.1.4}
$$

This representation is often a better starting point than the classical polynomial form for numerical applications. Moreover, in this matrix setting it is possible to develop a fast version of the Euclidean algorithm, where the quadratic cost of the classical algorithm is reduced to $\mathcal{O}(n \log^2 n)$. For more details on the classical and fast Euclidean algorithm see [19, 1, 24, 26].

3.1.2. Generalizations

Polynomial remainder sequences (PRS) are a generalization of the Euclidean algorithm, where each step is written as

$$a_i r_{i-1}(x) = q_i(x) r_i(x) + b_i r_{i+1}(x), \qquad i \geq 1$$

with $q_i(x), r_i(x) \in \mathbb{C}[x]$ and $a_i, b_i \in \mathbb{C}$. Different choices for a_i, b_i produce different sequences $\{r_i(x)\}$, but the final result, *i.e.* the polynomial $r_{s-1}(x)$ such that $r_s(x) = 0$, is always the GCD of $u(x)$ and $v(x)$, up to a scalar factor. We list here some common choices for a_i and b_i.

- If $a_i = b_i = 1$ for every $i > 1$, we obtain the classical Euclidean algorithm.
- Let $l(r_i)$ be the leading coefficient of $r_i(x)$ and $\delta_{i-1} = \deg r_{j-1} - \deg r_i > 0$. Then the *pseudo-division PRS* is obtained by setting

$$a_i = l(r_i)^{\delta_{i-1}+1}, \qquad b_i = 1, \qquad i > 1.$$

- The choice

$$u_i = l(r_i)^{\delta_{i-1}+1}, \qquad b_i = (-1)^{\delta_{i-1}} l(r_j)\beta_i, \qquad i > 1,$$

where the β_i's are defined recursively by

$$\beta_1 = 1, \qquad \beta_i = \frac{l(r_i)^{\delta_{i-1}}}{\beta_{i-1}^{\delta_{i-1}-1}},$$

gives the *subresultant PRS*. This particular sequence is often used in exact GCD computations, since it strikes a balance between coefficient growth (which might be exponential in the Euclidean algorithm) and complexity. See e.g. [94] for more details.

3.2. Stabilized numerical versions

One might think of adapting the Euclidean algorithm to the approximate case by simply converting all "small" coefficients into 0 after each polynomial division (where "small" coefficients are coefficients whose absolute value is smaller than a fixed threshold). Unfortunately, this naive method is highly unstable, as was already observed in [118].

Example 3.2.1. (Taken from [11].) Let

$$u(x) = x^4 + x^3 + (1 - \eta)x^2 + \eta x + 1,$$
$$v(x) = x^3 - x^2 + 3x - 2.$$

The first step of the Euclidean algorithm applied to $u(x)$ and $v(x)$ gives $r_1(x) = \eta x^2 + (\eta - 4)x + 5$. If η is below the fixed threshold, then one can eliminate terms of $r_1(x)$ and obtain a correct approximate GCD. But if η, though small, should happen to be just above the threshold, then a second division step should be performed, introducing significant numerical errors.

Example 3.2.2. ([118]) Another example of instability of the Euclidean algorithm is given by the polynomials

$$u(x) = x^4 + x + 1,$$
$$v(x) = x^3 - \eta x,$$

for which the first step of the Euclidean algorithm yields $r_1(x) = \eta x^2 + x + 1$.

It is possible, however, to design stabilized versions of the Euclidean algorithm. A first step in this direction is found in [100], where the authors introduce a normalization rule on remainders. The associated polynomial sequence is defined as follows:

- compute q_i as the quotient of r_{i-1} divided by r_i;
- set $r_{i-1} = q_i r_i + \max\{1, \|q_i\|_\infty\} \cdot r_{i+1}$.

This modification also allows to estabilish a relationship between ϵ and the distance between roots of the two polynomials.

An application of PRS with normalization to the computation of approximate polynomial GCD is also discussed in [116].

Another way to achieve better stability properties is a careful choice for the termination criterion. The problem is addressed in [73] and [74]. The analysis therein is based on the representations

$$r_i = s_j^{(i)} r_j + s_{j-1}^{(i)} r_{j+1}, \qquad j > i \geq 1 \qquad (3.2.1)$$

for the polynomials $r_i(x)$ generated by the Euclidean algorithm. The polynomials $s_j^{(i)}$ satisfy the recurrence relations

$$s_j^{(i)} = q_j s_{j-1}^{(i)} + s_{j-2}^{(i)}, \qquad j > i, \quad \text{with} \quad s_{i-1}^{(i)} = 0, \ s_i^{(i)} = 1, \quad (3.2.2)$$

which allow to compute the $s_j^{(1)}$'s and the $s_j^{(2)}$'s. Then it follows from (3.2.1) that the termination criterion

$$\|s_{j-1}^{(i)} r_{j+1}\| \leq \epsilon, \qquad i = 1, 2 \qquad (3.2.3)$$

estabilishes r_j as an ϵ-GCD of u and v. The Euclidean algorithm modified with this termination criterion is called *stabilized Euclidean algorithm.*

It must be pointed out, however, that both stabilized versions of the Euclidean algorithm presented in this section – and, more generally, any variant of the Euclidean algorithm – offer no guarantee that an ϵ-GCD has been found; indeed, they only return a common ϵ-divisor whose degree might be lower that the actual ϵ-GCD degree. A counterexample can be found in [50].

3.3. Relative primality

Beckermann and Labahn study in [10] and [11] the problem of approximate polynomial coprimeness (compare Problem 1.6.3) and provide criteria obtained via the Euclidean algorithm or making use of the Sylvester matrix.

Definition 3.3.1. For

$$u = \sum_{i=0}^{n} u_i z^i, \; v = \sum_{i=0}^{m} v_i z^i \in \mathbb{C}[z]$$

let

$$\begin{aligned} \epsilon(u, v) \;\; = \;\; & \inf \big\{ \|(u - \hat{u}, v - \hat{v})\|_1 : \\ & (\hat{u}, \hat{v}) \text{ have a common root, } \deg \hat{u} \leq n, \deg \hat{v} \leq m \big\}. \end{aligned}$$

In other words, any polynomials $a(z)$ and $b(z)$ satisfying $\|(u - a, v - b)\|_1 \leq \epsilon < \epsilon(u, v)$ and the above degree restriction are coprime.

The goal is to compute simple and approximately sharp lower bounds for $\epsilon(u, v)$.

A possible approach exploits the properties of the Sylvester matrix $S(u, v)$. For the purposes of this section, $S(u, v)$ denotes the transpose of the matrix defined in (2.5.1).

The following result is well known:

Lemma 3.3.2. *For any two polynomials u and v we have:*

$$\epsilon(u, v) \geq \frac{1}{\|S(u, v)^{-1}\|_1}.$$

Proof. It follows from the definition of $S(u, v)$ and from the choice of matrix norms that

$$\|S(u, v)\|_1 = \max\left\{\sum_{j=0}^{n} |u_j|, \sum_{j=0}^{m} |v_j|\right\} = \|(u, v)\|_1.$$

A theorem by Gastinel and Kahan (see [70], Theorem 6.5) states that the relative distance between a given matrix and the nearest singular one is the reciprocal of the condition number. In our case, this result implies that

$$
\begin{aligned}
\epsilon(u, v) &= \inf\{\|S(u, v) - S(\hat{u}, \hat{v})\|_1 : S(\hat{u}, \hat{v}) \text{ is singular}\} \\
&\geq \min\{\|S(u, v) - A\|_1 : B \text{ is singular}\} \\
&= \frac{1}{\|S(u, v)^{-1}\|_1}.
\end{aligned}
$$

$\square$

The proof of Lemma 3.3.2 shows that the quantity $\|(u, v)\|_1/\epsilon(u, v)$ may be considered as a structured condition number of $S(u, v)$ in the class of Sylvester matrices, with respect to the 1-norm. Indeed, Lemma 3.3.2 states that if we apply a perturbation of magnitude ϵ to the coefficients of u and v, where ϵ is smaller than the reciprocal of the norm of the inverse of the Sylvester matrix, then we still have relatively prime polynomials. This is essentially the 1-norm counterpart to the discussion of the SVD of the Sylvester matrix found in [41] (compare also [49] and [84]). The goal here, however, is to find a lower bound on $\epsilon(u, v)$ that

- is easier and cheaper to compute than the SVD,
- exploits the Sylvester structure (whereas the SVD does not take advantage of the structure),
- is larger than $1/\|S(u, v)^{-1}\|$.

It is a well known fact that u and v are coprime if and only if there exist polynomials a and b, with $\deg(a) < n$ and $\deg(b) < m$, such that

$$u \cdot b + v \cdot a = 1. \tag{3.3.1}$$

Equation (3.3.1) can also be written in terms of the Sylvester matrix as

$$S(u, v) \cdot \begin{bmatrix} \mathbf{b} \\ \mathbf{a} \end{bmatrix} = \begin{bmatrix} 1 \\ 0 \\ \vdots \\ 0 \end{bmatrix}.$$

This formula shows how the existence of a and b, and therefore the coprimeness of u and v, are strictly related to the existence of $S(u, v)^{-1}$, and in particular to its first column, as it is further emphasized by the following inversion formula:

Lemma 3.3.3. *Let*

$$f(z) = f_{-1}z^{-1} + \cdots + f_{1-m-n}z^{1-m-n} = \frac{a(z)}{u(z)} + \mathcal{O}(z^{-m-n})_{z \to \infty}.$$

Then $S(u, v)$ is invertible and its inverse is given by

$$S(u, v)^{-1} = A \cdot F,$$

where

$$
A = \begin{bmatrix}
b_0 & 0 & \cdots & \cdots & \cdots & 0 & v_0 & 0 & \cdots & \cdots & \cdots & 0 \\
\vdots & \ddots & \ddots & & & \vdots & \vdots & \ddots & \ddots & & & \vdots \\
b_{m-1} & \cdots & b_0 & 0 & \cdots & 0 & v_{m-1} & \cdots & v_0 & 0 & \cdots & 0 \\
a_0 & 0 & \cdots & \cdots & \cdots & 0 & -u_0 & 0 & \cdots & \cdots & \cdots & 0 \\
\vdots & \ddots & \ddots & & & \vdots & \vdots & \ddots & \ddots & & & \vdots \\
a_{n-1} & \cdots & a_0 & 0 & \cdots & 0 & -u_{n-1} & \cdots & -u_0 & 0 & \cdots & 0
\end{bmatrix},
$$

$$
F = \begin{bmatrix}
1 & 0 & \cdots & & & 0 \\
0 & \ddots & & \ddots & & \vdots \\
\vdots & & \ddots & & \ddots & 0 \\
0 & \cdots & & 0 & & 1 \\
0 & f_{-1} & & \cdots & & f_{1-m-n} \\
\vdots & & \ddots & & \ddots & \vdots \\
0 & & \cdots & & 0 & f_{-1}
\end{bmatrix}.
$$

Remark 3.3.4. A similar inversion formula can be derived for the Bezoutian matrix, yielding:

$$
B(u, v)^{-1} = \begin{bmatrix}
f_{-n} & f_{-n-1} & \cdots & f_{1-2n} \\
f_{1-n} & f_{-n} & \cdots & f_{2-2n} \\
\vdots & \vdots & & \vdots \\
f_{-1} & f_{-2} & \cdots & f_{-n}
\end{bmatrix}.
$$

What has been said so far suggests a numerical parameter that can be used to determine the relative coprimeness of two polynomials.

Theorem 3.3.5. *If the polynomials a and b satisfy (3.3.1) and have degrees at most $m - 1$ and $n - 1$, respectively, then*

$$\left\| \begin{bmatrix} b \\ a \end{bmatrix} \right\|_1 \leq \| S(u, v)^{-1} \|_1 \leq \left\| \begin{bmatrix} b \\ a \end{bmatrix} \right\|_1 + 2\| f \|_1 \|(u, v)\|_1. \quad (3.3.2)$$

Proof. The second inequality follows directly from the inversion formula of Lemma 3.3.3. The same inversion formula shows that the coefficients of a and b define the first column of $\| S(u, v)^{-1} \|_1$; hence the first inequality. $\qquad\square$

Observe that one may obtain similar results using the reversed polynomials

$$\underline{u}(z) = z^n \cdot u(1/z), \quad \underline{v}(z) = z^m \cdot v(1/z)$$

rather than u and v. Let us look at the solutions to the equation

$$\underline{u}(z) \cdot \tilde{b}(z) + \underline{v}(z) \cdot \tilde{a}(z) = 1. \quad (3.3.3)$$

If we take $\underline{a}(z) = z^{n-1} \cdot \tilde{a}(1/z)$ and $\underline{b}(z) = z^{m-1} \cdot \tilde{b}(1/z)$, then the above equation can be rewritten as

$$u(z) \cdot \underline{b}(z) + v(z) \cdot \underline{a}(z) = z^{m+n-1}, \quad (3.3.4)$$

that is,

$$S(u, v) \cdot \begin{bmatrix} \underline{b} \\ \underline{a} \end{bmatrix} = \begin{bmatrix} 0 \\ \vdots \\ 0 \\ 1 \end{bmatrix}.$$

In other words, the pair $(\underline{b}, \underline{a})$ is a $(n - 1, m - 1)$-Padé approximation for the rational function $-v(z)/u(z)$.

Now, define

$$\kappa = \left\| \begin{bmatrix} b & \underline{b} \\ a & \underline{a} \end{bmatrix} \right\|_1 = \max \left\{ \left\| \begin{bmatrix} b \\ a \end{bmatrix} \right\|_1, \left\| \begin{bmatrix} \underline{b} \\ \underline{a} \end{bmatrix} \right\|_1 \right\}.$$

Corollary 3.3.6. *The following inequalities hold:*

$$\kappa \leq \| S(u, v)^{-1} \|_1 \leq \kappa + 2 \cdot \| f \|_1 \cdot \|(u, v)\|_1, \quad (3.3.5)$$

where $\| f \|_1 = \| \underline{b} \cdot a - \underline{a} \cdot b \|_1$. Moreover, $\| f \|_1 \leq \kappa^2$.

Proof. The inequalities in (3.3.5) are a consequence of Theorem 3.3.5. As for $\|f\|_1$, we have:

$$f(z) - z^{1-m-n} \cdot [\underline{b}(z) \cdot a(z) - \underline{a}(z) \cdot b(z)]$$

$$= \frac{a(z) - u(z) \cdot z^{1-m-n} \cdot [\underline{b}(z) \cdot a(z) - \underline{a}(z) \cdot b(z)]}{u(z)} + \mathcal{O}(z^{-m-n})_{z \to \infty}$$

$$= \frac{z^{1-m-n} \cdot [v(z) \cdot \underline{a}(z) \cdot a(z) + u(z) \cdot \underline{a}(z) \cdot b(z)]}{u(z)} + \mathcal{O}(z^{-m-n})_{z \to \infty}$$

$$= z^{1-m-n} \cdot \frac{a(z)}{u(z)} + \mathcal{O}(z^{-m-n})_{z \to \infty}$$

$$= \mathcal{O}(z^{-m-n})_{z \to \infty},$$

and therefore

$$f(z) = z^{1-m-n} \cdot [\underline{b}(z) \cdot a(z) - \underline{a}(z) \cdot b(z)]. \qquad \square$$

From (3.3.5) we can immediately deduce that $1/(\kappa + 2 \cdot \|f\|_1 \cdot \|(u, v)\|_1)$ is a lower bound for $\|S(u, v)^{-1}\|_1$, and hence for $\epsilon(u, v)$. Unfortunately, this bound might potentially be of the order of $1/\kappa^2$, and therefore be too small to be useful. It can be proved, however, that κ itself, rather than $1/(\kappa + 2 \cdot \|f\|_1 \cdot \|(u, v)\|_1)$, is a suitable coprimeness parameter:

Theorem 3.3.7. (Corollary 4.4 in [10]) *The following inequality holds:*

$$\epsilon(a, b) \geq \frac{1}{\kappa}.$$

More on this topic is found in the next section.

3.4. A look-ahead method

In [11], the authors employ a modified version of the Euclidean algorithm (called *algorithm COPRIME*) to compute the coprimeness parameter κ defined in the previous section. The algorithm COPRIME is part of a Maple[1] package for approximate polynomial computations, called SNAP (Symbolic-Numeric Algorithms for Polynomials); see [75] for a description.

Recall from Section 3.1 that the Euclidean algorithm can be put into a matrix polynomial framework as

$$[u, v] \cdot U = [r_j, r_{j+1}],$$

---

[1] Maple is a registered trademark of Waterloo Maple Software.

where, if k is such that

$$\deg u > \deg v \geq \deg r_j = n - k > \deg r_{j+1},$$

then the elements of the 2×2 matrix polynomial U have degree bounds given by

$$\deg U \leq \begin{bmatrix} n - m + k - 1 & n - m + k \\ k - 1 & k \end{bmatrix}.$$

A matrix U as above is called a *unimodular reduction* of order k. It is also desirable that U be *scaled*, that is, both its columns have a norm between $1/2$ and 1 (see [11] for more details on scaling). Moreover, $\underline{U}(x)$ is an *associated vector* of order k if

$$[u(x), v(x)] \cdot \underline{U}(x) = x^{m+k-1} + \tilde{c}(z),$$

where

$$\deg \tilde{c} \leq n - k - 1, \qquad \deg \underline{U} \leq \begin{bmatrix} m - n + k - 1 \\ k - 1 \end{bmatrix}.$$

Lemma 3.4.1. *The following facts hold:*

 (i) *Any unimodular reduction has a polynomial inverse.*
 (ii) *The polynomials $u(x)$ and $v(x)$ are coprime $\Leftrightarrow$ there exists a unimodular reduction $U^{(n)}$ of order n $\Leftrightarrow$ there exists a unique associated vector $\underline{U}^{(n)}$ of order n.*
 (iii) *Solutions of (3.3.3) and (3.3.4) are obtained from the first column of $U^{(n)}$ and from $\underline{U}^{(n)}$. Furthermore,*

$$\epsilon(u, v) \geq \min \left\{ \frac{|\tilde{u}(0)|}{\|U^{(n)} \cdot [1, 0]^T\|}, \frac{1}{\|\underline{U}^{(n)}\|} \right\}.$$

Lemma 3.4.1 shows that solutions for (3.3.3) and (3.3.4) and a lower bound on $\epsilon(u, v)$ can be obtained by determining an exact unimodular reduction of order n together with its associated vector. When using finite arithmetic, however, numerical counterparts of these quantities are computed, so one must show that Lemma 3.4.1, or a similar result, still holds.

As suggested in [28], a numerical unimodular reduction of order n can be determined by recurrence in terms of unimodular reductions of lower order, and the same holds for the associated vectors. In other words, we want to successively construct numerical unimodular reductions of order

k, together with associate vectors, for increasing values of $k \in \mathcal{A} \subseteq \{1, 2, \ldots, n\}$:

$$[u^{(k)}, v^{(k)}] \leftarrow [u, v] \cdot U^{(k)}, \qquad x^{m+k-1} + c^{(k)} \leftarrow [u, v] \cdot \underline{U}^{(k)}.$$

Assume that a unimodular reduction of order k is known and a stepsize s has been chosen; then a unimodular reduction of order $k + s$ is computed as

$$U^{(k+s)} \leftarrow U^{(k)} \cdot U^{(k,k+s)}, \qquad \underline{U}^{(k+s)} \leftarrow x^s \cdot \underline{U}^{(k)} - U^{(k)} \cdot \underline{U}^{(k,k+s)},$$

with initializations

$$U^{(0)} = \begin{bmatrix} 1 & 0 \\ 0 & 1 \end{bmatrix}, \qquad \underline{U}^{(0)} = \begin{bmatrix} 0 \\ 0 \end{bmatrix}.$$

The transition factor $U^{(k,k+s)}$ is a 2×2 matrix polynomial computed by constructing a numerical unimodular reduction of order s of $[u^{(k)}, v^{(k)}]$:

$$[u^{(k+s)}, v^{(k+s)}] \leftarrow [u^{(k)}, v^{(k)}] \cdot U^{(k,k+s)}.$$

The crucial point in this algorithm is the choice of the stepsize s. A small s is advantageous from the point of view of computational cost, but in some cases it might be better to choose a larger s in order to "jump" over singular or ill-conditioned subproblems, thus ensuring good stability properties. For this reason this is labeled as a "look-ahead" algorithm: before solving each subproblem, the method checks whether it is well conditioned and, if this is not the case, the stepsize is increased, thus avoiding the numerically unstable part of the computation. Such ill-conditioned subproblems are encountered, for instance, in Examples 3.2.1 and 3.2.2.

The computation of the $U^{(k)}$'s goes as follows:

- Initialization: set $k = 0$ and $\mathcal{A} = \emptyset$.
- For $s = 1, 2, \ldots$ do:

  - Compute $U^{(k,k+s)}$;
  - If $|\det U^{(k+s)}(0)| > \epsilon$ and $\|\underline{U}^{(k+s)}\| < 1/\epsilon$ for a fixed tolerance ϵ, exit loop.

- Set $k \leftarrow k + s$, $\mathcal{A} \leftarrow \mathcal{A} \cup \{k\}$ and compute a unimodular reduction of order k.
- If $k = n$, exit algorithm.

The computational cost of this algorithm is typically quadratic in the degrees of the polynomials, since the stepsizes turn out to be seldom larger than 3. In pathological cases where $s = n$, however, the number of arithmetic operations needed increases to $\mathcal{O}(n^3)$.

The results provided by the algorithm COPRIME may be used to compute approximate GCDs.

Since computations are done using finite precision arithmetic, there are residual error polynomials $\alpha^{(k)}$ and $\beta^{(k)}$ such that

$$(u, v) = U^{(k)} \cdot (u^{(k)}, v^{(k)}) + (\alpha^{(k)}, \beta^{(k)})$$

with $\deg u^{(k)} = n - k > \deg v^{(k)}$. Let $\rho_l(u, v)$ be the minimum of the set of all products $\|(u, v)\| \cdot \|(g_u, g_v)^T\|$ (the polynomial 1-norm is being used here), where the polynomial pair (g_u, g_v) is such that $\deg g_u < l$, $\deg g_v < l$ and

$$z^m u(z) g_u(z) + z^n v(z) g_v(z) = z^{n+m+l-1} + \mathcal{O}(z^{n+m-1}).$$

Then the following result holds (see [75]):

Theorem 3.4.2. *Let $U^{(k)}$ be the last well-behaved unimodular reduction computed by the algorithm* COPRIME. *Then*

(a) *if $\|v^{(k)}\| + \|(\alpha^{(k)}, \beta^{(k)})\| < \epsilon|\det U^{(k)}(0)|/12$, with $0 < \epsilon < 1/6$, then $u^{(k)}$ is a quasi-GCD with precision ϵ;*
(b) *if $2\|v^{(k)}\| + (2 + 4\rho_{m-n+2k}(u, v)) \cdot \|(\alpha^{(k)}, \beta^{(k)})\| \le \epsilon|\det U^{(k)}(0)|$ and $|\det U^{(k)}(0)| > 4\rho_{m-n+2k}(u, v) \cdot \epsilon$, then $u^{(k)}$ is an ϵ-GCD.*

Theorem 3.4.2 is used in the SNAP package to compute quasi-GCDs and ϵ-GCDs.

3.5. Padé approximation

Let $a(x) = \sum_{i=0}^{+\infty} a_i x^i$ be a formal power series. Given nonnegative integers k and l, a (k, l)-th *Padé approximation* for $a(x)$ is a pair of polynomials $s(x)$ and $t(x)$ such that

$$s(x) - a(x)t(x) = 0 \qquad \bmod x^{k+l+1},$$
$$\deg s(x) \le k,$$
$$\deg t(x) \le l.$$

The polynomials $s(x)$ and $t(x)$ are uniquely defined, up to scaling by common factors or divisors.

Let us see now how the notion of Padé approximation is related to the polynomial GCD ([19],[103]). Assume that polynomials $u(x)$ and $v(x)$ are given. If $w(x)$ and $z(x)$ are polynomials of the smallest degree such that

$$w(x)v(x) = z(x)u(x),$$

then $g(x) = \text{GCD}(u(x), v(x))$ is found as

$$g(x) = \frac{u(x)}{w(x)} = \frac{v(x)}{z(x)}.$$

The (k, l)-th Padé approximation of two given polynomials can be computed via the extended Euclidean algorithm (see [24], [19]), with a computational cost of $\mathcal{O}((k + l) \log^2(k + l))$. These considerations easily give a GCD algorithm and can be extended to the heuristic computation of a common approximate divisor, which, however, will suffer from the difficulties related to the application of the Euclidean algorithm to approximate polynomial computations.

Padé approximations can also be computed by using methods specifically designed for Hankel and Bézout-structured matrices; such techniques have the advantages of enabling a better numerical control and providing an upper bound to the degree of the approximate GCD.

3.6. Euclidean algorithm and factorization of the Bezoutian

The topic of this section is the relationship between the Euclidean algorithm applied to a pair of polynomials $u(x)$ and $v(x)$ and the LU factorization of the associated Bézout or Hankel matrix. The main result essentially states that the entries of the triangular matrices of the block triangular factorization of $H(u, v)$ or $JB(u, v)J$ yield the coefficients of all the polynomials generated by the Euclidean scheme (see the works by Bini and Gemignani [15, 16, 59] and the book [19] for a more detailed presentation).

The following proposition explains how the polynomials generated by the Euclidean algorithm can be obtained through Schur complements of the permuted Bezoutian:

Proposition 3.6.1. *Let $u(x)$ and $v(x)$ be monic coprime polynomials of degrees respectively n and m, with $n > m$. Let H be the associated semi-infinite Hankel matrix as in Section 2.7 and let $0 = m_0 < m_1 < \cdots < m_l = n$ be integers such that $\det H_{m_i} \neq 0$ for $i = 0, \ldots, l$ and $\det H_k = 0$ otherwise.*

Then $\det(JB(u, v)J)_{m_i} \neq 0$ for $i = 1, \ldots, l$ and $\det(JB(u, v)J)_{m_j} = 0$ for $j \neq m_i, i = 1, \ldots, l$. Moreover, if S_{m_i} is the Schur complement of

$(JB(u, v)J)_{m_i}$ in $JB(u, v)J$, *the following relation holds:*

$$S_{m_i} = (-1)^i J B(r_i, r_{i+1}) J,$$

where $\{r_i(x)\}_{i=1,...,l}$ *is the polynomial remainder sequence obtained by applying the Euclidean scheme to* $u(x)$ *and* $v(x)$.

As for the triangular factorization of the Bezoutian, we have:

Proposition 3.6.2. *Under the hypothesis of Proposition 3.6.1, set* $Y_{m_i} = (JB(u, v)J)_{m_i}$. *Then*

$$JB(u, v)J = \begin{pmatrix} Y_{m_i} & G^T \\ G & W \end{pmatrix}$$

$$= \begin{pmatrix} I & 0 \\ GY_{m_i}^{-1} & I \end{pmatrix} \begin{pmatrix} Y_{m_i} & 0 \\ 0 & S_{m_i} \end{pmatrix} \begin{pmatrix} I & Y_{m_i}^{-1}G^T \\ 0 & I \end{pmatrix}$$

$$= \begin{pmatrix} T & 0 \\ GY_{m_i}^{-1}T & I \end{pmatrix} \begin{pmatrix} JB(u^{(i)}, v^{(i)})J & 0 \\ 0 & (-1)^i JB(r_i, r_{i+1})J \end{pmatrix}.$$

$$\cdot \begin{pmatrix} T^T & T^T Y_{m_i}^{-1}G^T \\ 0 & I \end{pmatrix},$$

where

$$S_{m_i} = W - GY_{m_i}^{-1}G^T = (-1)^i JB(r_i, r_{i+1})J,$$

$$T = \begin{pmatrix} u_n & & 0 \\ \vdots & \ddots & \\ u_{n-m_i+1} & \cdots & u_n \end{pmatrix} (B(u^{(i)}, v^{(i)})^{-1})J.$$

Therefore, the coefficient vector of the $(i + 1)$-h remainder $r_{i+1}(x)$ can be expressed as

$$\mathbf{r}_{i+1} = J(W - GY_{m_i}^{-1}G^T)J\mathbf{e}^{n-m_i},$$

that is, $\mathbf{r}_{i+1}$ can be computed by solving an $m_i \times m_i$ system with Hankel-like matrix Y_{m_i}.

Observe that the assumption in Proposition 3.6.1 that $u(x)$ and $v(x)$ are coprime is no loss of generality. Indeed, if $r_L(x) = \text{GCD}(u, v)$ were a polynomial of positive degree, then the Euclidean remainder sequence $\{r_i(x)\}$ can be recovered from the sequence $\{\tilde{r}_i(x)\}$ generated by the Euclidean algorithm applied to $u(x)/r_L(x)$ and $v(x)/r_L(x)$.

Chapter 4
Matrix factorization and approximate GCDs

This chapter describes some methods for approximate GCD computation which are based upon a rank-revealing factorization, of a resultant matrix. Such factorizations are useful for approximate rank determination, and therefore to determine bounds on the degree of an approximate GCD; a set of coefficients is then derived either from the factorization itself, or using other techniques. The results given by factorization-based methods are sometimes significantly improved by a refinement stage.

4.1. Approximate rank and the SVD

The singular value decomposition applied to the Sylvester matrix of two polynomials $u(x)$ and $v(x)$ has been often used in literature for investigating the degree of an approximate GCD, mainly because it is the main tool in numerical analysis for deciding the rank of a matrix in presence of perturbations.

The classical reference for a discussion of the SVD is [62]. We recall here the definition and some basic properties that will later prove useful.

Theorem 4.1.1. *Let A be a complex $m \times n$ matrix. Then there exist unitary matrices $U \in \mathbb{C}^{m \times m}$ and $V \in \mathbb{C}^{n \times n}$ and an $m \times n$ diagonal matrix $\Sigma = \mathrm{diag}\,(\sigma_1, \sigma_2, \ldots, \sigma_p)$, with $p = \min\{m, n\}$, such that*

$$A = U\Sigma V^H, \tag{4.1.1}$$

where

$$\sigma_1 \geq \sigma_2 \geq \cdots \geq \sigma_p \geq 0.$$

The factorization (4.1.1) is called the *singular value decomposition* of A and the σ_i's are the *singular values*. The columns $u_1, \ldots, u_m$ of U are the *left singular vectors* of A, whereas the columns $v_1, \ldots, v_m$ of V are the *right singular vectors*.

The SVD reveals a great deal of information on a matrix. If $A = U\Sigma V^H$ as in (4.1.1) and r is an index such that

$$\sigma_1 \geq \cdots \geq \sigma_r > \sigma_{r+1} = \cdots = 0,$$

then

$$\text{rank}(A) = r,$$
$$\ker(A) = \text{span}\{v_{r+1}, \ldots, v_n\},$$
$$\text{range}(A) = \text{span}\{u_1, \ldots, u_r\}.$$

Moreover, A can be written as

$$A = \sum_{i=1}^{r} \sigma_i u_i v_i^H.$$

Theorem 4.1.2. *With the above notation, if $k < r = \text{rank}(A)$ and*

$$A_k = \sum_{i=1}^{k} \sigma_i u_i v_i^H,$$

then

$$\min_{\text{rank}(B)=k} \|A - B\|_2 = \|A - A_k\|_2 = \sigma_{k+1}.$$

In other words, Theorem 4.1.2 says that the distance between A and the nearest matrix of rank k is σ_{k+1}.

This is a crucial result for the computation of the numerical or the approximate rank of a matrix, which in turn is fundamental in GCD computation. Indeed, from a numerical point of view it makes little sense to say that a matrix A is exactly rank deficient (i.e. some of its singular values are exactly zero), whereas a better approach is to regard A as *numerically rank deficient* if it is very close to an exactly rank deficient matrix (i.e. some of the singular values of A are very small). So the *numerical rank* of A is r if the last $n - r$ singular values of A are smaller than $\mu \|A\|_2$, where μ is the machine precision.

A very similar reasoning applies if there is an uncertainty ϵ on the entries of A (for example because A has been obtained from laboratory experiments). In this case it makes sense to compute the ϵ-*rank* of A: we say that $r_\epsilon = \text{rank}(A, \epsilon)$ if

$$\sigma_1 \geq \cdots \geq \sigma_{r_\epsilon} > \epsilon \geq \sigma_{r_\epsilon+1} \geq \cdots \geq \sigma_p.$$

4.1.1. The SVD of Sylvester and Bézout matrices

Recall from Theorems 2.5.2 and 2.6.7 that if $S \in \mathbb{C}^{N \times N}$ is the Sylvester matrix associated with polynomials $u(x)$ and $v(x)$ and rank$(S) = k$, then $N - k = \deg \mathrm{GCD}(u, v)$, and a similar property is true for $B(u, v)$, the Bezoutian of $u(x)$ and $v(x)$.

It is clear that the notion of approximate rank is crucial when transposing this result to the approximate case. Indeed, the degree of an ϵ-GCD of $u(x)$ and $v(x)$ is $N - \mathrm{rank}(S, \eta)$, for a proper choice of η. In many cases, though, finding the correct ϵ-GCD degree through the SVD of $S(u, v)$ or $B(u, v)$ is not easy. However, we compute in the following two lemmas some useful bounds that can be readily obtained. We assume here that $\|u(x)\|_2 = \|v(x)\|_2 = 1$.

Lemma 4.1.3. *Let $u(x)$ and $v(x)$ be polynomials of degree respectively n and m, with $n \geq m$. Suppose that both $u(x)$ and $v(x)$ have unitary 2-norm. If $\epsilon > 0$ is given, and $\hat{u}(x)$ and $\hat{v}(x)$ are such that $\deg \hat{u} = n$, $\deg \hat{v} = m$, $\|u(x) - \hat{u}(x)\|_2 \leq \epsilon$, $\|v(x) - \hat{v}(x)\|_2 \leq \epsilon$, then up to a first order approximation:*

$$\|B(u, v) - B(\hat{u}, \hat{v})\|_2 < 4n\epsilon. \tag{4.1.2}$$

Proof. Let $a(x) = \hat{u}(x) - u(x)$ and $b(x) = \hat{v}(x) - v(x)$, where $\|a(x)\|_2 \leq \epsilon$ and $\|b(x)\|_2 \leq \epsilon$. From the linearity of the Bezoutian it follows that

$$B(\hat{u}, \hat{v}) = B(u, v) + B(u, b) + B(a, v) + B(a, b).$$

Since we are interested in a first order approximation, we have

$$B(u, v) - B(\hat{u}, \hat{v}) \dot{=} B(u, b) + B(a, v)$$

and therefore

$$\|B(u, v) - B(\hat{u}, \hat{v})\|_2 \leq \|B(u, b)\|_2 + \|B(a, v)\|_2.$$

From Lemma 4.1.3 we have

$$\|B(u, b)\|_2 \leq 2n\|u(x)\|_2\|b(x)\|_2 \leq 2n\epsilon$$
$$\|B(a, v)\|_2 \leq 2n\|a(x)\|_2\|v(x)\|_2 \leq 2n\epsilon$$

and the thesis immediately follows. $\qquad\qquad\square$

Lemma 4.1.4. *Assume that the polynomial $u(x)$ and $v(x)$ have an ϵ-GCD of degree k.*

(a) *Let $\sigma_1 \geq \cdots \geq \sigma_N$ be the singular values of $S(u, v)$. Then*

$$\sigma_{N-k+1} \leq \epsilon\sqrt{N}.$$

(b) *Let $\sigma_1 \geq \cdots \geq \sigma_n$ be the singular values of $B(u, v)$. Then*

$$\sigma_{n-k+1} \leq 4n\epsilon + \mathcal{O}(\epsilon^2).$$

Proof. Let $\hat{u}(x)$ and $\hat{v}(x)$ be polynomials of the same degrees as $u(x)$ and $v(x)$, such that $\|u - \hat{u}\|_2 \leq \epsilon$, $\|v - \hat{v}\|_2 \leq \epsilon$, and having an exact GCD of degree k. The matrix $S(\hat{u}, \hat{v})$ has rank $N - k$, so it follows from Theorem 4.1.2 that $\sigma_{N-k+1} \leq \|S(\hat{u}, \hat{v})\|_2$. Moreover, by Lemma 2.5.6 we have $\|S(u, v) - S(\hat{u}, \hat{v})\|_2 = \|S(u - \hat{u}, v - \hat{v})\|_2 \leq \epsilon\sqrt{N}$, so (a) is proved.

Analogously, we have $\sigma_{n-k+1} \leq \|B(\hat{u}, \hat{v})\|_2$. It follows from Lemma 4.1.3 that $\|B(\hat{u}, \hat{v})\|_2 \leq 4n\epsilon$, thus yielding (b). □

In [40], Corless *et al.* discuss the SVD of the Sylvester matrix in the context of approximate GCD finding. Part of their analysis relies on the following result :

Lemma 4.1.5. *Let $\sigma_1, \ldots, \sigma_N$ be the singular values of S. If*

$$\sigma_1 \geq \sigma_2 \geq \cdots \geq \sigma_k > \epsilon\sqrt{N} > \epsilon \geq \sigma_{k+1} \geq \cdots \geq \sigma_N,$$

and if g is a common divisor of $u + \Delta u$ and $v + \Delta v$ with $\deg g \geq N - k + 1$, then either $\|\Delta u\|_2 > \epsilon$ or $\|\Delta v\|_2 > \epsilon$ (or both).

Proof. Let $S + E$ be a Sylvester matrix of rank $N - k + 1$ which is nearest to S. Then $\|E\|_2 \geq \sigma_k$. Observe that $S(u + \Delta u, v + \Delta v)$ is also a matrix of rank $N - k + 1$. Thus $S(\Delta u, \Delta v) = S(u + \Delta u, v + \Delta v) - S(u, v)$ is such that $\|S(\Delta u, \Delta v)\|_2 \geq \|E\|_2 \geq \sigma_k > \epsilon\sqrt{N}$. If, by contradiction, we had $\|\Delta u\|_2 \leq \epsilon$ and $\|\Delta v\|_2 \leq \epsilon$, then by Lemma 2.5.6 we would have $\|S(\Delta u, \Delta v)\|_2 \leq \epsilon\sqrt{N}$. □

The algorithm for approximate GCD proposed in [41] takes as input the two polynomials $u(x)$ and $v(x)$ and a tolerance $\epsilon > 0$; the output is an $\hat{\epsilon}$-GCD $d(x)$, where $\hat{\epsilon}$ (which is larger than ϵ) is computed *a posteriori*. Notice that this method addresses a problem which is slightly different from the standard approximate GCD problem, since the perturbation is not bounded *a priori*. It should also be pointed out that the discussion in [40] focuses more on the GCD degree rather than on the computation of a set of coefficients.

The algorithm goes as follows:

- Form $S = S(u, v)$;
- Compute the SVD $S = U \Sigma V^T$;
- Find the maximum k such that $\sigma_k > \epsilon \sqrt{N}$ and $\sigma_{k+1} \leq \epsilon$ (if all singular values are larger than $\epsilon \sqrt{N}$ then set $d = 1$; if there is no such gap in the singular values, then report failure). The index k is the declared rank of S, and the degree of d is $\deg(u) - k$;
- Compute the coefficients of $d(x)$ (using for example a modified Euclidean algorithm, or Gaussian elimination on the range of S, or optimization techniques, or root-finding methods).

If $\hat{u}(x)$ and $\hat{v}(x)$ are computed by solving the associated least-squares problem, then $\hat{\epsilon}$ can be defined as the maximum between $\|u - \hat{u}\|_2$ and $\|v - \hat{v}\|_2$.

It is pointed out in [50] that this method might sometimes output an $\hat{\epsilon}$-divisor which does not have maximum degree. Indeed, it might happen that, once k has been found and an approximate divisor $d(x)$ of degree k has been computed, the actual perturbation $\hat{\epsilon}$ is large enough to allow an $\hat{\epsilon}$-divisor of degree larger than k.

An interesting feature of the approach adopted in [40] is the emphasis on the importance of the existence of a suitable gap in the singular values of $S(u, v)$. Such a gap suggests a "natural" choice for the approximate GCD degree. So what the algorithm actually computes is an approximate divisor whose degree is the most "reasonable", if perturbations larger than ϵ are allowed (of course such perturbations still have to be kept as small as possible).

It may happen, however, that there is no clear separation amongst singular values. This is often the case, for example, for polynomials of large degree. In such a situation, there is no "natural" choice for the approximate GCD degree, and this is why the algorithm simply reports failure. The standard ϵ-GCD problem, on the contrary, can still be solved.

Example 4.1.6. (Taken from [41]) Let

$$u(x) = x^5 + 5.503x^4 + 9.765x^3 + 7.647x^2 + 2.762x + 0.37725,$$
$$v(x) = x^4 - 2.993x^3 - 0.7745x^2 + 2.0070x + 0.7605.$$

The singular values of $S(u, v)$ are approximately

$$23.1, \ 14.6, \ 7.62, \ 4.68, \ 3.59, \ 2.72, \ 1.11, \ 1.41 \times 10^{-4}, \ 6.11 \times 10^{-6}.$$

Inspection of the singular values shows that the most reasonable choice for the approximate GCD degree is $k = 2$.

A candidate GCD is then computed, yielding

$$d(x) = x^2 + 1.007x + 0.2534.$$

The minimum possible perturbation to the polynomials that allows exact division by $d(x)$ turns out to have norm about 1.6×10^{-4}, which is only slightly larger than the eighth singular value.

4.1.2. The SVD of subresultants

More bounds on the degree of an approximate GCD can be obtained through the SVD of subresultant matrices (see Section 2.5 for definition and notation).

Following [50], denote by τ_r the smallest singular value of S_r, and by $\gamma_{n-1} \geq \cdots \geq \gamma_0 \geq 0$ the last m singular values of S_0 (this just means renaming the last m singular values of $S(u, v)$ as $\gamma_k = \sigma_{m+n-k}$). It is readily seen that $\tau_0 = \sigma_{m-n}$ and $\tau_m > 0$.

The following results are proved in [49] and [50].

Lemma 4.1.7. *With the previous notation, we have*

$$\tau_m \geq \tau_{m-1} \geq \cdots \geq \tau_1 \geq \tau_0 = \gamma_0.$$

Theorem 4.1.8. *Each of the bounds*

$$\epsilon \leq \frac{\tau_r}{\sqrt{m+n-2r}}$$

and

$$\epsilon \leq \frac{\gamma_r}{\sqrt{m+n}}$$

implies $\deg \epsilon\text{-GCD}(u, v) \leq r$.

Theorem 4.1.9. *Suppose that*

$$\tau_{r-1} \leq \epsilon \leq \tau_r / \sqrt{2} \quad \text{for} \quad \epsilon \leq 1, \quad 1 \leq r \leq m \leq n$$

and

$$\tau_r > \tau_{r-1} 2^{2n+m-2r}.$$

If, moreover,

$$\left(1 + \frac{2 + \tau_r^2}{\tau_r - \tau_{r-1}}\right)^{n-1} \left(1 + \frac{2^{2n+m-2r}}{\tau_r - \tau_{r-1} 2^{2n+m-2r}}\right) \tau_{r-1} \leq \epsilon,$$

then the degree of an ϵ-GCD *of* $u(x)$ *and* $v(x)$ *is equal to* r.

Theorem 4.1.9 is a certification theorem, that is, it has the merit of providing conditions under which the degree of an ϵ-GCD is guaranteed, whereas other results usually only provide either upper or lower bounds. Its limitation lies in the exponential factor present in the required gap, which reduces the practical significance of the theorem, even though these are worst-case bounds, which should not be attained in most practical applications.

The following corollary gives a simpler version of the gap.

Corollary 4.1.10. *Suppose that* $\tau_{r-1} \leq \epsilon \leq \tau_r/\sqrt{2}$, $\tau_r \leq 23/25$ *and* $1 \leq r \leq m \leq n$. *If*

$$\tau_{r-1} \leq 2^{-5n-3} \tau_r^{n+2} \epsilon,$$

then the degree of an ϵ-GCD of $u(x)$ and $v(x)$ is equal to r.

The results presented in this section, combined, if necessary, with more bounds such as the ones given either by the Euclidean algorithm or by the SVD, can be used to devise several variants of algorithms for the computation of an approximate GCD; see [49] for details.

4.1.3. Statistical approaches

We mention here that, in order to reduce the computational effort required by the SVD, some authors have proposed algorithms for approximate polynomial GCD that exploit fast SVD-related techniques based on a statistical treatment of errors.

The algorithm proposed in [124], for example, appears in the setting of a system identification application and assumes that the polynomial coefficients are Gaussian random variables; the GCD coefficients are estimated through a maximum-likelihood procedure.

In the context of adaptive control theory, Zarowski ([141]) proposes to use a condition estimator ([20]) to determine the smallest singular values of a resultant matrix and subsequently estimate the approximate GCD degree through the approximate rank estimator examined in [140]. This estimator treats errors on singular values as independent random variables, with probability density functions chosen in order to obtain a good tradeoff between a tractable and simple model and a physically accurate one.

Winkler and Allan also worked along similar lines to develop a polynomial GCD- and rootfinder; they employ techniques based on the principle of MDL (minimum description length) to estimate the approximate rank of a Sylvester matrix without need for a user-defined tolerance.

4.2. An SVD-based algorithm

The algorithm for the computation of polynomial GCD proposed by Z. Zeng in [144] basically relies on rank determination and computation of the null space of Sylvester subresultant matrices, followed by iterative refinement. The definition of approximate GCD used is the one presented in Section 1.1.3 (AGCD), where the normalization vector $\mathbf{r}$ is defined by the tentative set of AGCD coefficients available before the refinement stage.

The rank-revealing technique employed by Zeng uses an iterative method to approximate the smallest singular value of the studied matrix; details can be found in [142]. For the case of Sylvester subresultants, the result is:

Lemma 4.2.1. *For a fixed $j \in \{0, 1, \ldots, n-1\}$, let $Q_j R_j = S_j(u, v)$ be a QR decomposition of $S_j(u, v)$, let ς_j be the smallest singular value of $S_j(u, v)$, with $\mathcal{Y}_j$ being the associated right singular subspace of R_j, and let $\tau_j = \|R_j\|_\infty$. For every initial vector $\mathbf{z}_0$ not orthogonal to $\mathcal{Y}_j$, the iteration*

$$\begin{cases} \mathbf{w}_i = \mathbf{z}_{i-1} - \left(\frac{2\tau_j \mathbf{z}_{i-1}^H}{R_j}\right)^\dagger \left(\frac{\tau_j \mathbf{z}_{i-1}^H \mathbf{z}_{i-1} - \tau}{R_j \mathbf{z}_i}\right) \\ \mathbf{z}_i = \frac{\mathbf{w}_i}{\|\mathbf{w}_i\|_2}, \qquad \sigma_i = \|R_j \mathbf{z}_i\|_2, \end{cases}$$

with $i = 1, 2, \ldots$ produces two sequences $\{\sigma_i\}$ and $\mathbf{z}_i$ such that:

(a) $\lim_{i \to \infty} \sigma_i = \varsigma_j$,

(b) $\lim_{i \to \infty} d_i = 0$, *where d_i is the distance between $\mathbf{z}_i$ and $\mathcal{Y}_j$,*

(c) *if ξ_j is the smallest singular value of R_j that is larger than ς_j, then*

$$|\sigma_{i+1} - \varsigma_j| \le \left(\frac{\varsigma_j}{\xi_j}\right)^2 |\sigma_i - \varsigma_j|$$

$$d_{i+1} \le \left(\frac{\varsigma_j}{\xi_j}\right)^2 d_i.$$

It follows from *(b)* that if $u(x)$ and $v(x)$ have a GCD of degree $k = n - j$, then $\dim(\mathcal{Y}_j) = 1$ and the sequence $\mathbf{z}_i$ converges to a vector in the null space of R_j.

A relationship between the tolerance ϵ and the degree of an ϵ-GCD is also provided: namely, if the degree of an ϵ-GCD is k, then

$$\varsigma_{n-k} \le \epsilon \sqrt{2(n - k + 1)}. \tag{4.2.1}$$

For a polynomial pair, the method works as follows. Starting with $j = 0$, consider the matrix $S_j(u, v)$ and compute its QR decomposition $S_j(u, v) =$

$Q_j R_j$. Apply the iteration of Lemma 4.2.1 to find a pair ς_j, $\mathbf{y}_j$. If ς_j satisfies (4.2.1), with $k = n - j$, then there is a possibility that the ϵ-GCD degree might be k. In order to check if this is the case, recall that the vector $\mathbf{y}_j$ contains the coefficients of tentative cofactors $p(x)$ and $q(x)$. Once $p(x)$ and $q(x)$ are retrieved, compute a tentative ϵ-GCD $g(x)$ via least squares division (see Section B.1.1) and refine the result using the Newton method applied to the system

$$\begin{cases} u(x) = g(x)p(x) \\ v(x) = g(x)q(x) \end{cases}$$

with an additional normalizing equation. If the obtained polynomial is an ϵ-GCD, the algorithm stops. Otherwise, set $j = j + 1$ and repeat the procedure.

Observe that, since $S_{j+1}(u, v)$ is obtained from $S_j(u, v)$ by adding two columns containing the coefficients of $u(x)$ and $v(x)$, the computational cost of the QR decomposition of $S_{j+1}(u, v)$ can be reduced using QR update techniques. The overall computational cost is cubic in the degrees of the input polynomials.

4.3. QR factorization of the Sylvester matrix

In [41], Corless *et al.* propose an algorithm for the computation of approximate polynomial GCD based upon Theorem 2.5.3. The obvious way to apply Theorem 2.5.3 to the approximate case consists in performing the QR decomposition of the Sylvester matrix associated with the given polynomials and taking as coefficients af an approximate GCD the entries of the last row of significant magnitude in the computed upper triangular factor.

The authors, though, point out that the straightforward application of Theorem 2.5.3 to the approximate case presents stability issues. While it is true that QR decomposition is backward stable if performed using, for instance, Givens rotations or Householder transformations, there is no similar bound on the forward error, which may grow quite large.

More precisely, let $\hat{R}$ be a computed upper triangular factor of a given matrix S; assume that the computation is performed using Givens or Householder transformations. Then there exists an orthogonal matrix $\hat{Q}$ such that

$$S + \Delta S = \hat{Q}\hat{R}, \tag{4.3.1}$$

$$\|\Delta S\|_F \leq \eta \|S\|_F, \quad \eta = O(\mu).$$

However, if R is the upper triangular factor in the exact QR decomposition of S, there is no guarantee that the forward error

$$\Delta R = \hat{R} - R$$

should have small norm.

Experimental evidence shows that, if S is the Sylvester matrix of polynomials $u(x)$ and $v(x)$, the forward error in the QR decomposition of S tends to grow large when $u(x)$ and $v(x)$ have large common roots. The authors of [41] give a formalization of this idea. In the following, $u(x)$ and $v(x)$ are assumed to have unitary 2-norm.

Proposition 4.3.1. *Let $u(x)$ and $v(x)$ be given univariate approximate polynomials. Let $\{x_i\}_{1 \leq i \leq k}$ be common roots (in the approximate sense) of $u(x)$ and $v(x)$ which lie inside the unit circle, i.e. $|x_i| \leq \rho < 1$ for all i. There may be other common roots lying outside the unit circle. Then the* QR *factorization of the Sylvester matrix of $u(x)$ and $v(x)$ reveals a factor of the approximate* GCD *of $u(x)$ and $v(x)$ that contains the zeros $\{x_i\}_{1 \leq i \leq k}$.*

Proof. Denote with $u + \Delta u$ and $v + \Delta v$ perturbed polynomials such that the x_i's are roots of the exact GCD of $u + \Delta u$ and $v + \Delta v$. Let S be the Sylvester matrix associated with $u + \Delta u$ and $v + \Delta v$. Let N be the null space N of S,which is parametrized by the x_i's as in Proposition 2.5.5. As above, let $S = QR$ be the exact QR decomposition of S and $\hat{R}$ the actually computed upper triangular factor; ΔS and $\hat{Q}$ are such that $S + \Delta S = \hat{Q}\hat{R}$. Then we have:

$$(S + \Delta S)N = \Delta S N = \hat{Q}\hat{R}N$$

and therefore

$$\hat{R}N = \hat{Q}^H \Delta S N. \qquad (4.3.2)$$

Equation (4.3.2) can be interpreted as polynomial evaluation on the common zeros of $u + \Delta u$ and $v + \Delta v$. In particular, the polynomial $r(x)$ arising from the last nonzero row of $\hat{R}$ is bounded in value, when evaluated at each x_i; this implies that the x_i's are pseudozeros of $r(x)$. Therefore $r(x)$ is close, in a dual norm, to the common approximate divisor of $u(x)$ and $v(x)$ that has the x_i's as roots. □

The above proposition ensures that the QR factoring method allows to find at least a common factor whose roots include all the small common roots of $u(x)$ and $v(x)$. The authors of [41] give the following explanation for the fact that large common roots might not be detected.

Let x be a real or complex number. It follows from (4.3.1) that

$$\left(\begin{bmatrix} x^{m-1} f \\ \vdots \\ f \\ x^{n-1} g \\ \vdots \\ g \end{bmatrix} + \Delta S \right) \begin{bmatrix} x^{n+m-1} \\ \vdots \\ x^{n-1} \\ \vdots \\ x \\ 1 \end{bmatrix} = \hat{Q} \hat{R} \begin{bmatrix} x^{n+m-1} \\ \vdots \\ x^{n-1} \\ \vdots \\ x \\ 1 \end{bmatrix}.$$

Let $\mathbf{x} = [x^{n+m-1} \ \ldots \ x \ 1]^T$ and recall that $\mathbf{x}$ belongs to the null space of S if and only if $x = x_i$ for some i. So the common roots of $u(x)$ and $v(x)$ will still correspond to the null space of $\hat{Q} \hat{R}$ if and only if the perturbation term $\Delta S \mathbf{x}$ can be neglected. With probability 1, ΔS is not a Sylvester matrix itself and hence $\mathbf{x}$ is not nearly in its null space. So it is reasonable to assume that $\| \Delta S \cdot \mathbf{x} \| \approx \| \Delta S \| \cdot \| \mathbf{x} \|$. This means that, if $|x| > 1$, the perturbation term may no longer be negligible.

A common factor associated with large common roots can be retrieved by applying the QR factoring technique to the reversed polynomials $\hat{u}(x) = x^n u(1/x)$ and $\hat{v}(x) = x^m v(1/x)$. So one can

- use the QR factorization to compute a common factor $d_1(x)$ containing common roots that lay inside the unit circle;
- subsequently apply the QR factoring method to the reversals of $u(x)/d_1(x)$ and $v(x)/d_2(x)$, obtaining another common factor $d_2(x)$;
- compute a GCD as $d(x) = d_1(x) d_2(x)$.

Difficulties might arise, though, when some common roots are very close to the unit circle. Indeed, if there is not a clear separation of common roots inside the unit circle from common roots that lie outside, it is not possible to correctly identify $d_1(x)$ and $d_2(x)$ from a gap in the magnitude of the rows of $\hat{R}$.

In order to overcome this obstacle, the authors of [41] propose to use Graeffe's root squaring technique. Let us recall the basic idea behind this classical technique. Given a polynomial

$$f(x) = f_n(x - \alpha_1)(x - \alpha_2) \ldots (x - \alpha_n),$$

define another polynomial $g(x)$ as follows:

$$\begin{aligned} g(x) &= (-1)^n f(-\sqrt{(x)}) f(\sqrt{x}) & (4.3.3) \\ &= (-1)^n f_n^2 (x - \alpha_1^2)(x - \alpha_2^2) \ldots (x - \alpha_n^2). & (4.3.4) \end{aligned}$$

Equation (4.3.4) shows that $g(x)$ is a polynomial whose roots are the squares of the roots of $f(x)$. The coefficients of $g(x)$ can be computed using the relation (4.3.3); the computation can be carried out using the fast Fourier transform in time $O(n \log n)$.

Each root-squaring step will improve the separation of the roots from the unit circle; but since it may also separate initially close complex roots, it is not advisable to perform more than two or three root-squaring steps.

In cases when even root-squaring fails to give a significant improvement, more sophisticated splitting techniques are required. Common roots which are close to the unit circle, though, may still represent a difficult case for this type of QR factoring approach.

This algorithm has been incorporated in Maple 9.

Chapter 5
Optimization approach

The optimization approach to the computation of an approximate GCD is based upon the following formulation of the problem:

Problem 5.0.2. Given the coefficients of polynomials $u(x) = \sum_{j=1}^{n} u_j x^j$ and $v(x) = \sum_{j=1}^{m} v_j x^j$ and a positive integer $k \leq n, m$, find the coefficients of polynomials $\hat{u}(x)$, $\hat{v}(x)$ and $g(x)$ of degrees respectively $\hat{n} \leq n$, $\hat{m} \leq m$ and k such that:

- $g(x)$ is a common divisor of $\hat{u}(x)$ and $\hat{v}(x)$;
- the perturbation norm

$$\eta = \|u - \hat{u}\|_2^2 + \|v - \hat{v}\|_2^2$$

is minimized over all the triples $(\hat{u}, \hat{v}, g)$. (Of course the problem can be reformulated using other polynomial norms).

The problem is well-posed. Indeed we have, for the 2-norm version (see [80]):

Theorem 5.0.3. *With the notation of Problem 5.0.2, given real (complex) polynomials $u(x)$, $v(x)$ and an integer k, there exist perturbed real (complex) polynomials $\hat{u}(x)$ and $\hat{v}(x)$ such that for all polynomials $\tilde{u}(x)$ and $\tilde{v}(x)$ with $\deg \tilde{u} \leq n$, $\deg \tilde{v}(x) \leq m$ and $\deg \mathrm{GCD}(\tilde{u}, \tilde{v}) \geq k$ we have*

$$\|\hat{u} - u\|_2^2 + \|\hat{v} - v\|_2^2 \leq \|\tilde{u} - u\|_2^2 + \|\tilde{v} - v\|_2^2.$$

In view of this formulation, solution techniques are focused on the minimization of the function η, with appropriate constraints and parametrizations.

Remark 5.0.4. Optimization methods usually require an initial guess among the input data. In other words, a tentative set of coefficients for an approximate common divisor of the prescribed degree should be provided in order to ensure that solution techniques for Problem 5.0.2 work properly. On the other hand, the accuracy of direct GCD methods (such as the

ones examined in Chapter 4) would often be greatly improved by iterative refinement. Therefore, direct methods and optimization techniques
can be used together in the design of effective algorithms for approximate
polynomial GCD.

For example, if an upper bound k_0 on the degree of an ϵ-GCD is
known, one can do the following:

- compute an approximate common divisor $g_0(x)$ of degree k_0;
- apply refinement;
- if $g_0(x)$ is an ϵ-divisor, then it is an ϵ-GCD;
  else decrease k_0 by 1 and repeat the procedure.

The quantity η – called the *strength* of the approximation in [83] – therefore provides a backward error for approximate GCD computations.

5.1. The case of degree one

Early suggestions for an optimization approach are found in [40] and
[100]. A detailed study is proposed by Karmarkar and Lakshman ([84]
and [85]), who examine the case $k = 1$. Their analysis goes as follows.

Let $u(x)$ and $v(x)$ be given complex polynomials as in Problem 5.0.2,
and suppose we perturb them so that the perturbed polynomials $\hat{u}(x)$ and
$\hat{v}(x)$ have a common factor $x - \alpha$. We are assuming here to work with
monic polynomials. We have

$$\hat{u}(x) = (x - \alpha)\left(\sum_{j=0}^{n-1} \phi_j x^j\right) = \sum_{j=0}^{n} (\phi_{j-1} - \alpha\phi_j)x^j,$$

$$\hat{v}(x) = (x - \alpha)\left(\sum_{j=0}^{m-1} \gamma_j x^j\right) = \sum_{j=0}^{m} (\gamma_{j-1} - \alpha\phi_j)x^j,$$

where $\phi_{-1} = \phi_n = 0$ and $\phi_{n-1} = 1$, and analogously $\gamma_{-1} = \gamma_m = 0$ and
$\gamma_{m-1} = 1$. The function $\eta = \|u - \hat{u}\|^2 + \|v - \hat{v}\|^2$ that is to be minimized
is then parametrized by α and can be written as

$$\eta = \mathbf{y}^H Q \mathbf{y} - (\mathbf{y}^H \mathbf{r} + \mathbf{r}^H \mathbf{y}) + \mathbf{s}, \tag{5.1.1}$$

where the entries of the vectors $\mathbf{y}$, $\mathbf{r}$ and $\mathbf{s}$ are linear functions of the
coefficients of $u(x)$, $v(x)$, $\hat{u}(x)$ and $\hat{v}(x)$, whereas Q is a tridiagonal
matrix that depends only on α and is Hermitian and positive definite for
every α.

For any choice of $\alpha \in \mathbb{C}$, the function η has a minimum at a stationary
point; let us denote the value of η at this stationary point as η_m. Then η_m
is a real-valued function of the complex variable $\alpha = a + ib$. We now
seek to minimize η_m with respect to α.

If η_m is seen as a function of the real variables a and b, then its stationary points are given by the equations

$$\frac{\partial \eta_m}{\partial a} = \frac{\partial \eta_m}{\partial b} = 0. \tag{5.1.2}$$

Observe that we are only interested in real solutions of (5.1.2). Let $\eta_m(a, b)$ attain its absolute minimum at the point (ζ, ξ); then $\alpha = \zeta + i\eta$ is a common root of the perturbed polynomials $\hat{u}(x)$ and $\hat{v}(x)$ that satisfy the minimum perturbation requirement. An estimate on the precision needed in the computation of the real roots of (5.1.2) is also provided, as well as a bound on the region of the $\mathbb{R}^2$-plane where such roots should lie if they represent a minimum for η_m.

The main steps of Karmarkar and Lakshman's algorithm are therefore as follows:

- Determine η as in (5.1.1);
- Compute the "symbolic minimum" η_m, along with rational functions in α for the coefficients of $\hat{u}(x)$ and $\hat{v}(x)$;
- Compute the real solutions of 5.1.2, with the prescribed precision and bounds;
- Among the solutions found, choose (ζ, ξ) where η_m assumes the minimum value;
- Set $\alpha = \zeta + i\xi$ and compute coefficients for $\hat{u}(x)$ and $\hat{v}(x)$.

These techniques can also be extended to the classical formulation of the approximate GCD problem or variations of it. An example is given by the problem of finding the nearest singular polynomial to a given one. This minimization problem was studied earlier by Hough ([72]), who used an approach based on Lagrange multipliers.

An interesting development of Karmarkar and Lakshman's approach can be found in [31] and [32]. The authors tackle the problem of minimizing the function η by introducing a subdivision algorithm, which is based on an inclusion and an exclusion test. The exclusion test relies on Taylor development of relevant polynomials to exclude the existence of a global minimum of the function within a given square in the complex plane. The inclusion test, on the contrary, uses Smale's α-theory to certify the existence and uniqueness of a solution in a square. Certification is therefore a crucial and attractive feature of this method.

5.2. Structured low-rank approximation

Let $u(x) = \sum_{i=0}^n u_i x^i$ and $v(x) = \sum_{i=0}^m v_i x^i$ be numerical polynomials. We have already pointed out in Section 2.5 that their Sylvester matrix $S(u, v)$ may give useful information on the approximate GCD of $u(x)$ and $v(x)$.

More precisely, let us suppose that $g(x) = \sum_{i=0}^{k} g_i x^i$ is an ϵ-GCD for a fixed tolerance ϵ. It follows from the definition of ϵ-GCD that there are polynomials $\hat{u}(x) = \sum_{i=0}^{n} \hat{u}_i x^i$ and $\hat{v}(x) = \sum_{i=0}^{m} \hat{v}_i x^i$ such that

- $g(x)$ is an exact divisor of $\hat{u}(x)$ and $\hat{v}(x)$;
- $\|u(x) - \hat{u}(x)\| \leq \epsilon$, $\|v(x) - \hat{v}(x)\| \leq \epsilon$, for a chosen polynomial norm.

Then it follows from Theorem 2.5.2 and Lemma 2.5.6 that:

- $S(\hat{u}, \hat{v})$ has rank $n + m - k$;
- $\|S(u, v) - S(\hat{u}, \hat{v})\| \leq K\epsilon$,

where K is a function of the norms and degrees of the polynomials and depends on the chosen vector and matrix norm. Notice that $S(u, v)$ will likely have full numerical rank, or at least a greater rank than $S(\hat{u}, \hat{v})$.

The converse is also true: if $\hat{S}$ is a Sylvester-structured matrix of rank $n + m - k$ and such that $\|S(u, v) - \hat{S}\| \leq K\epsilon$, then the polynomials associated to $\hat{S}$ play the role of $\hat{u}(x)$ and $\hat{v}(x)$ above.

So, the idea behind the use of structured low rank approximation for computing the approximate polynomial GCD is: given ϵ and $S(u, v)$, find a Sylvester-structured matrix $\hat{S}$ of lowest possible rank, such that $\|S(u, v) - \hat{S}\| \leq K\epsilon$. Since the Sylvester structure is linear, one may equivalently say: find a Sylvester-structured perturbation matrix S_ϵ such that $S(u, v) - S_\epsilon$ is of lowest rank and $\|S_\epsilon\| \leq K\epsilon$.

The Sylvester structure of the perturbation, and therefore of $\hat{S}$, is a crucial issue here, because it allows to retrieve the polynomials $\hat{u}(x)$ and $\hat{v}(x)$ and control the distances $\|u(x) - \hat{u}(x)\|$ and $\|v(x) - \hat{v}(x)\|$. If we were not interested in structure, an ordinary SVD, for instance, could provide a low rank approximation of $S(u, v)$ in 2-norm, but the obtained low-rank matrix would be unstructured and would give no information on the coefficients of an approximate divisor.

Several algorithms for structured low rank approximation are found in the literature. They usually require k among the input data and return a structured matrix of rank k; in terms of polynomials, this means that what such algorithms actually compute is an approximate divisor of given degree k. Structured low rank approximation techniques, therefore, do not immediately give an approximate GCD, but they can nevertheless be used in approximate GCD algorithms.

5.2.1. Lift-and-project algorithm

The following iterative method is presented in [35]. Assume that polynomials $u(x)$ and $v(x)$ are given as above and let k be an integer (a tentative degree for an approximate GCD or divisor). Denote with

- $\mathcal{R}_k$ the set of $(m + n) \times (m + n)$-matrices of rank k,

- $\mathcal{S}$ the set of $(m+n) \times (m+n)$-matrices having a Sylvester structure.

Both $\mathcal{R}_k$ and $\mathcal{S}$ can be seen as algebraic varieties in $\mathbb{R}^{(m+n)\times(m+n)}$. The lift-and-project method consists in alternating projections between $\mathcal{R}_k$ and $\mathcal{S}$, so that the rank constraint and the structure constraint are satisfied alternatively, while reducing the distance in-between (see Figure 5.1).

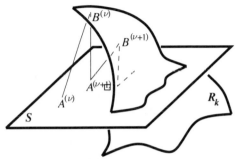

Figure 5.1. Lift-and-project algorithm.

Algorithm
Start with an arbitrary matrix $A^{(0)} = A \in \mathcal{S}$ and iterate the following two steps for $\nu = 0, 1, \ldots$ until convergence:

1. *(lift)* compute the matrix $B^{(\nu)}$ of rank-k which is nearest to A^ν;
2. *(project)* compute the orthogonal projection $A^{(\nu+1)}$ of $B^{(\nu)}$ on $\mathcal{S}$.

The lift can be performed using the truncated SVD. Given $A^{(\nu)} \in \mathcal{S}$, compute its SVD

$$A^{(\nu)} = U^{(\nu)} \Sigma^{(\nu)} V^{(\nu)T},$$

where

$$\Sigma^{(\nu)} = \begin{bmatrix} \sigma_1^{(\nu)} & & & \\ & \sigma_2^{(\nu)} & & \\ & & \ddots & \\ & & & \sigma_{(m+n)}^{(\nu)} \end{bmatrix}$$

and $\sigma_1^{(\nu)}, \ldots, \sigma_{(m+n)}^{(\nu)}$ are the singular values of $A^{(\nu)} \in \mathcal{S}$. Replace $\Sigma^{(\nu)}$ with

$$\Sigma_t^{(\nu)} = \begin{bmatrix} \sigma_1^{(\nu)} & & & & & \\ & \ddots & & & & \\ & & \sigma_k^{(\nu)} & & & \\ & & & 0 & & \\ & & & & \ddots & \\ & & & & & 0 \end{bmatrix}$$

and define

$$B^{(\nu)} = U^{(\nu)} \Sigma_t^{(\nu)} V^{(\nu)T}.$$

The matrices in the sequence $\{A^{\nu}\}$ obtained from the algorithm do not necessarily have rank k. But notice that the following property holds:

$$\|A^{\nu+1} - B^{\nu+1}\|_F \leq \|A^{\nu+1} - B^{\nu}\|_F \leq \|A^{\nu} - B^{\nu}\|_F, \qquad (5.2.1)$$

so the described algorithm is a descent algorithm. Property 5.2.1 ensures that, if the matrices A^{ν} are distinct, then the sequence converges to a structured matrix of rank k, that will be called $P_k(A)$.

It should be remarked that the above iteration, if it converges, does give a matrix of rank k in S, but there is no guarantee that this should be the closest rank-k Sylvester matrix to A. An optimization stage is needed, where the following function should be minimized:

$$f(T) = \|A - P_k(T)\|.$$

5.2.2. Structured total least norm

Structured low rank approximation can also be achieved using structured total least norm (STLN) techniques. In [149], the authors modify a method proposed in [107] to compute a structure preserving low rank approximation of a Sylvester matrix.

As above, let $S(u, v)$ be the Sylvester matrix of given polynomials $u(x) = \sum_{i=0}^{n} u_i x^i$ and $v(x) = \sum_{i=0}^{m} v_i x^i$.

Lemma 5.2.1. *Partition $S(u, v)^T$ as $[B \quad A]$, where B is given by the first k columns of $S(u, v)^T$ and A by the last $m + n - k$. Then*

$$\text{rank}(S(u, v)) \leq m + n - k \Leftrightarrow AX = B \text{ has a solution.}$$

Proof. If $AX = B$ has a solution, then the columns of B are in the range of A, so $S(u, v)$ cannot have rank greater than $m + n - k$.

Conversely, suppose $\text{rank}(S(u, v)) \leq m + n - k$; we seek a solution to $AX = B$. In order to have a better understanding of what a solution would look like, multiply both sides of $AX = B$ on the left by $[x^{m+n-1}, x^{m+n} \ldots, x, 1]$, so that the system is expressed in terms of polynomials. What is obtained is

$$[x^{m-k-1} f(x) \quad x^{m-k-2} f(x) \ldots f(x) \quad x^{n-1} g(x) \ldots g(x)]X$$
$$= [x^{m-1} f(x) \quad x^{m-2} f(x) \ldots x^{m-k} f(x)].$$

Therefore it is possible to find a solution X if there are polynomials $\{p_i(x)\}$ and $\{q_i(x)\}$ such that

$$\deg p_i(x) \le m - k - 1,$$
$$\deg q_i(x) \le n - 1,$$
$$x^{m-1} f(x) = p_i(x) f(x) + q_i(x) g(x), \quad \text{for} \quad i = 1, \dots k. \quad (5.2.2)$$

Now, if $\operatorname{rank}(S(u, v)) \le m + n - k$, then it follows from Theorem 2.5.2 that $u(x)$ and $v(x)$ have a GCD $g(x)$ of degree $d \ge k$. Let $u_1(x) = u(x)/g(x)$ and $v_1(x) = v(x)/g(x)$ be the cofactors and denote by $a_i(x)$ and $b_i(x)$ the quotient and the remainder of the division of x^{m-i} by $v_1(x)$:

$$x^{m-i} = a_i(x) v_1(x) + b_i(x).$$

Then

$$p_i(x) = b_i(x),$$
$$q_i(x) = a_i(x) u_1(x)$$

are solutions of (5.2.2).   □

Lemma 5.2.1 allows to reformulate the problem as follows: find a STLN solution to the overdetermined system

$$AX \approx B, \quad (5.2.3)$$

i.e., find structured perturbation matrices of minimum norm F and E such that the columns of $B + F$ belong to the range of $A + E$.
    Let us first examine the case $k = 1$. The system (5.2.3) becomes

$$Ax \approx b,$$

where $A \in \mathbb{C}^{(m+n) \times (m+n-1)}$, $b \in \mathbb{C}^{m+n}$ and $[b \quad A]$ is a (transposed) Sylvester matrix. A structured perturbation $[f \quad E]$ can be represented by a vector $\eta \in \mathbb{C}^{m+n+2}$ as follows:

$$[f \quad E] = \begin{pmatrix} \eta_1 & 0 & \cdots & 0 & 0 & \eta_{n+2} & 0 & \cdots & 0 & 0 \\ \eta_2 & \eta_1 & \cdots & 0 & 0 & \eta_{n+3} & \eta_{n+2} & \cdots & 0 & 0 \\ \vdots & \vdots & \cdots & \vdots & \vdots & \vdots & \vdots & \cdots & \vdots & \vdots \\ 0 & 0 & \cdots & \eta_{n+1} & \eta_n & 0 & 0 & \cdots & \eta_{m+n+2} & \eta_{m+n+1} \\ 0 & 0 & \cdots & 0 & \eta_{n+1} & 0 & 0 & \cdots & 0 & \eta_{m+n+2} \end{pmatrix}.$$

Observe that $f = P_1 \eta$, where

$$P_1 = \begin{pmatrix} I_{n+1} & 0 \\ 0 & 0 \end{pmatrix} \in \mathbb{R}^{(m+n) \times (m+n+2)}.$$

The structured residual is

$$\mathbf{r}(\eta, \mathbf{x}) = \mathbf{b} + \mathbf{f} - (A + E)\mathbf{x},$$

so our problem now consists in computing

$$\min_{\eta, \mathbf{x}} \|\eta\|_2 \quad \text{with} \quad \mathbf{r} = \mathbf{0}. \tag{5.2.4}$$

A way to solve (5.2.4) involves the use of the penalty method, seeking

$$\min_{\eta, \mathbf{x}} \left\| \begin{array}{c} w\mathbf{r}(\eta, \mathbf{x}) \\ \eta \end{array} \right\|_2, \tag{5.2.5}$$

where w is a large penalty value, for example 10^{10}. The computation of (5.2.5) is simplified by using a linear approximation to $\mathbf{r}(\eta, \mathbf{x})$, with respect to perturbations of η and $\mathbf{x}$:

$$\mathbf{r}(\eta + \Delta\eta, \mathbf{x} + \Delta\mathbf{x}) = \mathbf{b} + P_1(\eta + \Delta\eta) - (A + E + \Delta E)(\mathbf{x} + \Delta\mathbf{x})$$
$$\doteq \mathbf{b} + P_1\eta - (A + E)\mathbf{x} + P_1\Delta\eta - (A + E)\Delta\mathbf{x} - \Delta E\mathbf{x}.$$

If $\mathbf{x} = [x_1\, x_2 \ldots x_{m+n-1}]^T$, define the Sylvester-structured matrix

$$Y_1 = \begin{pmatrix} 0 & & & & x_m & & \\ x_1 & \ddots & & & x_{m+1} & \ddots & \\ \vdots & \ddots & 0 & & \vdots & \ddots & x_m \\ x_{m-1} & & x_1 & x_{m+n-1} & & & x_{m+1} \\ & \ddots & \vdots & & \ddots & & \vdots \\ & & x_{m-1} & & & & x_{m+n-1} \end{pmatrix},$$

so that

$$Y_1\eta = E\mathbf{x}.$$

Now, problem (5.2.5) can be reformulated as

$$\min_{\Delta\mathbf{x}, \Delta\eta} \left\| \begin{pmatrix} w(Y_1 - P_1) & w(A + E) \\ I_{m+n+2} & 0 \end{pmatrix} \begin{pmatrix} \Delta\eta \\ \Delta\mathbf{x} \end{pmatrix} + \begin{pmatrix} -w\mathbf{r} \\ \eta \end{pmatrix} \right\|_2, \tag{5.2.6}$$

which is an ordinary least squares problem.

The case $k > 1$ can be solved in a similar way. By stacking the columns of $X = [\mathbf{x}_1 \ldots \mathbf{x}_k]$ and $B = [\mathbf{b}_1 \ldots \mathbf{b}_k]$, the system (5.2.3) can be written as

$$\begin{pmatrix} A & 0 & \cdots & 0 \\ 0 & A & \cdots & 0 \\ \vdots & & \ddots & \vdots \\ 0 & \cdots & 0 & A \end{pmatrix} \begin{pmatrix} \mathbf{x}_1 \\ \mathbf{x}_2 \\ \vdots \\ \mathbf{x}_k \end{pmatrix} \approx \begin{pmatrix} \mathbf{b}_1 \\ \mathbf{b}_2 \\ \vdots \\ \mathbf{b}_k \end{pmatrix} \tag{5.2.7}$$

and can therefore be solved as in the case $k = 1$. Notice, though, that the size of the system (5.2.7) increases quadratically with k, so the algorithm may suffer from efficiency and stability problems. For this reason, when $k > 1$ it is advisable to work with a submatrix of $S(u, v)$ rather than the whole matrix. More precisely, choose a suitable $(m + n - k + 1) \times (m + n - 2k + 2)$ subresultant S_k of $S(u, v)$ and apply the techniques described above to find a structured total least squares solution of the system

$$A_k \mathbf{x} \approx \mathbf{b_k}$$

obtained from the partition $S_k = [\mathbf{b_k} \quad A_k]$.

A similar approach has recently been taken by Winkler and collaborators ([134, 136, 137]), who consider polynomials $u(x)$ and $\alpha v(x)$, with α a properly chosen parameter, and solve a constrained least squares problem of the type (5.2.4) via the QR decomposition.

5.3. The divisor - quotient iteration

In [33], Chin *et al.* examine several strategies to solve Problem 5.0.2.

Using the same notation as in Problem 5.0.2, denote the perturbation polynomials as

$$\Delta u(x) = \hat{u}(x) - u(x),$$
$$\Delta v(x) = \hat{v}(x) - v(x).$$

We then have

$$g(x)p(x) = u(x) + \Delta u(x), \quad (5.3.1)$$
$$g(x)q(x) = v(x) + \Delta v(x) \quad (5.3.2)$$

and the function to be minimized can be expressed as

$$\eta(g, p, q) = \|\Delta u\|^2 + \|\Delta v\|^2. \quad (5.3.3)$$

It is assumed that an initial guess for $g(x)$ is available.

In matrix notation, Equation (5.3.1) can be written as

$$
\begin{bmatrix} \Delta u_0 \\ \Delta u_1 \\ \vdots \\ \vdots \\ \Delta u_n \end{bmatrix}
=
\begin{bmatrix}
g_0 & & & \\
g_1 & g_0 & & \\
\vdots & g_1 & \ddots & \\
g_k & & \ddots & g_0 \\
& g_k & & g_1 \\
& & \ddots & \vdots \\
& & & g_k
\end{bmatrix}
\begin{bmatrix} p_0 \\ p_1 \\ \vdots \\ p_{n-k} \end{bmatrix}
-
\begin{bmatrix} u_0 \\ u_1 \\ \vdots \\ \vdots \\ u_n \end{bmatrix}
$$

and the same can be done for Equation (5.3.2). In a more compact form, we have

$$\Delta \mathbf{u} = C_g \mathbf{p} - \mathbf{u},$$
$$\Delta \mathbf{v} = C_g \mathbf{q} - \mathbf{v},$$

where C_g is a convolution matrix of appropriate size whose entries are given by the coefficients of $g(x)$. Since polynomial multiplication is commutative, we may also write

$$\Delta \mathbf{u} = C_p \mathbf{g} - \mathbf{u},$$
$$\Delta \mathbf{v} = C_q \mathbf{g} - \mathbf{v}.$$

Then (5.3.3) becomes

$$\eta(g, p, q) = \|C_g \mathbf{p} - \mathbf{u}\|^2 + \|C_g \mathbf{q} - \mathbf{v}\|^2 \tag{5.3.4}$$

or

$$\eta(g, p, q) = \|C_p \mathbf{g} - \mathbf{u}\|^2 + \|C_q \mathbf{g} - \mathbf{v}\|^2. \tag{5.3.5}$$

A minimization strategy for $\eta(g, p, q)$ – either in the form (5.3.4) or (5.3.5)– is sought.

One possibility, of course, is to employ a standard minimization technique such as the Levenberg-Marquardt method. But other approaches are possible, which also allow to divide the problem into smaller sub-problems.

The following technique, which was first proposed in [39], is known as *divisor-quotient iteration*. Define three polynomial sequences

$$\{g^{(j)}, p^{(j+1)}, q^{(j+1)}\}_{j=0,1,2,\dots} \tag{5.3.6}$$

as follows: Take $g^{(0)}$ as the given initial guess for $\mathbf{g}$ and compute

- $p^{(j+1)}$ and $q^{(j+1)}$ such that $\eta(g^{(j)}, p^{(j+1)}, q^{(j+1)})$ is minimized (with fixed $g^{(j)}$),
- $g^{(j+1)}$ such that $\eta(g^{(j+1)}, p^{(j+1)}, q^{(j+1)})$ is minimized (with fixed $p^{(j+1)}$ and $q^{(j+1)}$).

Then one has

Proposition 5.3.1. *The sequence of function values generated by the divisor-quotient iteration*

$$\{\eta(g^{(j)}, p^{(j)}, q^{(j)})\}_{j=1,2,\dots}$$

is convergent.

Under additional hypotheses, the polynomials sequences are also convergent. For example:

Proposition 5.3.2. *Assume that the divisor obtained at each iteration is monic (i.e., $g_k^{(j)} = 1$ for all j). Then each sequence in (5.3.6) is convergent.*

The main advantage of the divisor-quotient method lies in the fact that each iteration is computationally less expensive than in a standard minimization method, because a potentially large problem has been reduced to solving three smaller linear least-squares systems. Besides, the method is quite simple to implement. The convergence rate, on the other hand, is at best linear, and may be very slow.

5.4. More optimization strategies

We mention here some more optimization techniques and variants of previously described approaches that have recently been applied in the context of approximate GCD computation.

5.4.1. Another STLN approach

In [95], the authors exploit existent structured total least squares methods to solve Problem 5.0.2. Assume that a tentative GCD degree has been chosen. Writing the minimization problem along the lines of (5.3.4) and (5.3.5) and performing analytical elimination of constraints yields

Theorem 5.4.1. *With the notation of Problem 5.0.2, the minimization of η is equivalent to finding*

$$\min_{g_0,\dots,g_{k-1}\in\mathbb{R}} \text{trace}\left([\mathbf{u}\quad\mathbf{v}]^T\left(\mathbf{I} - C_g(C_g^T C_g)^{-1}C_g^T\right)[\mathbf{u}\quad\mathbf{v}]\right),$$

where $C_g \in \mathbb{R}^{(n+1)\times(n-k+1)}$ is a convolution matrix associated with the polynomial $g(x)$ and can be seen as a lower triangular banded Toeplitz matrix with first column equal to

$$[g_0, g_1, \dots, g_{k-1}, 1, 0, \dots, 0]^T.$$

It is assumed here that the polynomials involved in the approximate GCD problem have real coefficients and the GCD is monic.

As a consequence of Theorem 5.4.1, the distance function to be minimized can now be written as

$$\tilde{\eta}(g) = \text{trace}\left([\mathbf{u}\quad\mathbf{v}]^T\left(\mathbf{I} - C_g(C_g^T C_g)^{-1}C_g^T\right)[\mathbf{u}\quad\mathbf{v}]\right). \tag{5.4.1}$$

This new minimization problem can be solved using any standard optimization algorithm, for example the BFGS quasi-Newton method. During this optimization stage, several evaluations of $\tilde{\eta}(g)$ must be performed. The most expensive operation required by the evaluation of $\tilde{\eta}(g)$ is the solution of a least squares system with matrix C_g. This computation can be performed cheaply either applying a fast QR factorization via the generalized Schur algorithm (see [79]), or solving the associated normal system of equations with total least squares methods (see [96] and [97]) that exploit the banded Toeplitz structure of the matrix $C_g^T C_g$.

5.4.2. Sums of squares relaxation

The application of this optimization technique to the approximate GCD problem has received a certain interest in the last years: see *e.g.* [90], [99] and references therein.

Sum of squares (SOS) relaxation is aimed at solving global optimization problems where the objective function (*i.e.*, the function to be minimized) is a multivariate polynomial or rational function. The main idea consists in approximating the objective function via a SOS. As the name says, a multivariate polynomial $p(X) \in \mathbb{R}[X]$ is a SOS if it can be written as $p(X) = \sum_{j=1}^{k} q_j(X)$, with $q_j(X) \in \mathbb{R}[X]$ for $j = 1, ..., k$. SOS polynomials form a convex cone. Moreover, they are nonnegative; the converse is true in the univariate and quadratic case.

SOS relaxation can be used in the context of approximate GCD computations by reformulating the GCD problem as minimization of a polynomial or rational function. For instance, the quantity η defined in Problem 5.0.2 can be written as a nonnegative multivariate polynomial, whose variables are the coefficients of the GCD and of the cofactors. Minimization of η corresponds to computation of an approximate GCD.

5.4.3. Gradient-projection methods

The paper [126] tackles the approximate GCD problem by focusing on the application of a modified Newton method derived from the gradient-projection method ([112]) for constrained optimization. With the notation of Problem 5.0.2, let $a(x)$ and $b(x)$ be polynomials of degrees $m - k$ and $n - k$, such that $a(x)\hat{u}(x) + b(x)\hat{v}(x) = 0$. The objective function here is the perturbation norm η, seen as a function of the coefficients of $\hat{u}(x)$, $\hat{v}(x)$, $a(x)$ and $b(x)$. An extra constraint is added to enforce the condition $\|a\|_2^2 + \|b\|_2^2 = 1$. The resulting algorithm, named GPGCD, applies gradient-projection-based optimization to compute the coefficients of $\hat{u}(x)$, $\hat{v}(x)$, $a(x)$ and $b(x)$; the GCD $g(x)$ is obtained by least-squares division. In the output, the perturbed polynomials $\hat{u}(x)$ and $\hat{v}(x)$ are replaced by $b(x) \cdot g(x)$ and $a(x) \cdot g(x)$, respectively, to ensure that $g(x)$ is an exact common factor.

Chapter 6
New factorization-based methods

This and the next chapter are devoted to the presentation of some new methods for the computation of an approximate polynomial GCD that rely on suitable factorizations of resultant matrices combined with iterative refinement.

We adopt here Definition 1.1.3 of ϵ-GCD, with the usual vector 2-norm and with normalization of the polynomials; for simplicity, we assume that the input polynomials are already normalized to unitary 2-norm. Normalization (also called *row equilibration* when the Sylvester matrix is used, see [41]) often improves numerical stability, in addition to simplifying some estimates needed in the algorithms. Nonetheless, the algorithms presented here can be modified to handle the non-normalized case, if required.

The problem we want to solve is formulated like Problem 1.6.1, which we recall below: given the coefficients of real or complex normalized polynomials $u(x) = \sum_{i=0}^{n} u_i x^i$ and $v(x) = \sum_{i=0}^{m} v_i x^i$ and a tolerance ϵ, compute an ϵ-GCD $g(x)$ for $u(x)$ and $v(x)$, along with perturbed polynomials $\hat{u}(x)$ and $\hat{v}(x)$ of the same degrees as $u(x)$ and $v(x)$, with $\|\hat{u} - u\|_2 \leq \epsilon$ and $\|\hat{v} - v\|_2 \leq \epsilon$, and such that $g(x)$ is their exact GCD, and cofactors $p(x)$ and $q(x)$.

The new methods here presented and analyzed in detail are three. The algorithm TdBez (see Section 6.5) is based on the tridiagonalization of the Bézout matrix and only works with real polynomials.

The algorithm PivQr relies on the QR factorization with pivoting of the Sylvester or Bézout matrix and on the solution of a linear system defined by a submatrix of the Sylvester matrix. This algorithm can be applied to either real or complex polynomials; if the input polynomials have real coefficients, then the ϵ-GCD and cofactors returned will be real, too. Both these methods have cubic complexity in the degrees of the polynomials.

The next chapter presents a fast algorithm with quadratic complexity, based upon the fast LU factorization of displacement structured matrices. Here the output polynomials are (in general) always complex, regardless of whether the input coefficients were real or complex. See also [12].

6.1. Real vs. complex coefficients

The problem of whether approximate GCDs should have real or complex coefficients needs to be further discussed. If the input polynomials are complex, clearly the approximate GCD will in general have complex coefficients. But what if the input polynomials are real?

In an exact setting, the monic GCD of a pair of real polynomials is always real. But this is no longer true in the approximate setting. It might happen that, for certain values of ϵ, the degree of an ϵ-GCD is higher if complex coefficients are allowed. Equivalently, in the formulation of Problem 5.0.2, it might happen that complex perturbations of the given real polynomials allow to reach a lower value for η than strictly real perturbations. See [80] for a discussion.

Example 6.1.1. (From [80]) Take $u(x) = x^2 + 1$ and $v(x) = x^2 + 2$. Assume we want to find an approximate GCD of degree 1 according to the optimization approach (*i.e.* as in Problem 5.0.2). Then it can be shown that an optimal complex solution (not unique) requires a perturbation of magnitude (in 2-norm) $\eta_1 = 0.1007615$. The optimal real solution gives $\eta_2 = 0.145898$.

The example can be adapted to the approach of Problem 1.6.1 by choosing ϵ such that $\eta_1 < \epsilon < \eta_2$ and comparing the degrees of complex and real ϵ-GCDs.

So we are left with the problem of defining which kind of output is desired. This really depends on the purpose for which the approximate GCD is computed. The fact that input polynomials ar real may be taken to mean that the computations are to be kept within the real field; besides, the use of complex arithmetic makes computations more expensive. On the other hand, allowing a complex approximate GCD gives better results in terms of nearness or degree.

It should be pointed out, however, that an algorithm that would ordinarily yield a real approximate GCD if the input is real (like PivQr) will give complex results (with related advantages) if tiny complex perturbations are introduced at some point in the algorithm.

6.2. Sylvester vs. Bézout matrices

The algorithms for approximate GCD presented in this chapter and in the next one are based upon factorizations of the Sylvester and Bézout matrices. In some cases the role of these matrices is interchangeable (*e.g.* when QR or fast LU factorizations are applied), whereas other algorithms exploit peculiar properties of one of the two types of matrices (*e.g.* symmetry of the Bezoutian in the TdBex algorithm). We point out here some differences between the numerical properties of the Sylvester and Bézout matrices; as a consequence, some algorithms may be better suited than others to certain types of GCD problems.

- Let $u(x)$ and $v(x)$ be polynomials of degrees n and m respectively, with $n \geq m$. Then the Bézout matrix $B(u, v)$ associated with $u(x)$ and $v(x)$ has smaller size than the corresponding Sylvester matrix $S(u, v)$; namely, $B(u, v)$ is an $n \times n$ matrix, whereas $S(u, v)$ is $(n + m) \times (n + m)$. This property of the Bézoutian is advantageous both from the point of view of computational complexity and of numerical stability.
- On the other hand, the computation of the Sylvester matrix is straightforward if the coefficients of the polynomials are given, whereas more effort is required for $B(u, v)$. Moreover, the form of the Bézout matrix implies that cancellation may potentially occur.
- $S(u, v)$ and $B(u, v)$ also display different behaviours with respect to the scaling of polynomials. Indeed, for any $\alpha, \beta \in \mathbb{C}$ we have:

$$B(\alpha u, \beta v) = \alpha \beta B(u, v),$$
$$S(\alpha u, \beta v) = \begin{pmatrix} \alpha I_m & 0 \\ 0 & \beta I_n \end{pmatrix} S(u, v).$$

This is due to the fact that the polynomial coefficients are decoupled in the Sylvester matrix and implies, among other things, different responses of the two matrices to the normalization of $u(x)$ and $v(x)$. See [133] for a discussion of this behaviour with respect to the problem of defining a condition number for resultant matrices.

6.3. QR Factorization

As seen in Section 4.3, the QR factorization of the Sylvester matrix is the core of a well-estabilished method for approximate GCD computations. In this section we further analyze it and extend it to other resultant matrices, and design a QR-based algorithm for the computation of an ϵ-GCD.

6.3.1. Stability of the QR factorization

It follows from Theorems 2.5.3 and 2.6.11 that, in order to compute the exact GCD of two polynomials $u(x)$ and $v(x)$, one can compute a QR factorization of $S(u, v)$ or $J B(u, v) J$ and take the entries of the last nonzero row of R as coefficients of a GCD.

The same method can be applied to the approximate case, provided that the notion of "zero rows" is replaced by "rows whose norm is lower than a fixed tolerance δ". QR decompositions can be performed via Householder transformations, which ensure backward stability.

The use of the QR decomposition for estimating the ϵ-GCD degree and possibly computing a set of coefficients has some advantages over the use of the SVD. Indeed, it almost entirely employs rational operations, and it is usually slightly (though not asymptotically) cheaper. Moreover, it leaves open the possibility of designing a fast algorithm which relies on the specific displacement structure. In fact, fast algorithms for the LU factorization of displacement structured matrices exist; the design of fast algorithms for the QR factorization is work in place.

Nevertheless, as pointed out in [41], there is no guarantee that the forward error on the triangular factor is small. Numerical experiments show that the GCD method based upon the QR factorization of $S(u, v)$ or $J B(u, v) J$ might miss large common roots of $u(x)$ and $v(x)$, thus yielding, instead of a GCD, a common factor whose degree is not the highest possible one. This difficulty seems to affect the factorization of the Sylvester matrix more than the Bezoutian, but it is present in both cases. In order to overcome this obstacle, we analyze how perturbations of a resultant matrix propagate throughout its factorization process.

Remark 6.3.1. It follows from Lemma 2.5.6 that

$$\|R\|_2 \leq \sqrt{m\|f(x)\|_2^2 + n\|g(x)\|_2^2}.$$

Let S be the Sylvester matrix of the (exact) polynomials $f(x)$ and $g(x)$ and $\tilde{S}$ the Sylvester matrix of the perturbed polynomials $\tilde{f}(x) = f(x) + f_\epsilon(x)$ and $\tilde{g}(x) = g(x) + g_\epsilon(x)$, where $\|f_\epsilon(x)\|_2 \leq \epsilon$ and $\|g_\epsilon(x)\|_2 \leq \epsilon$. We have

$$\tilde{S} = S + E,$$

where E is the Sylvester matrix of $f_\epsilon(x)$ and $g_\epsilon(x)$.

It follows from Lemma 2.5.6 that

$$\|E\|_2 \leq \eta := \epsilon \sqrt{n + m}.$$

Let $S = QR$ be the QR decomposition of S. Then we have:

$$
\begin{aligned}
\tilde{S} &= S + E \\
&= QR + E \\
&= Q(R + Q^T E) \\
&= Q(R + F),
\end{aligned}
$$

where we have set $F = Q^T E$. Now, if $R + F = \hat{Q}\hat{R}$ is the QR decomposition of $R + F$, it follows that $\tilde{S} = (Q\hat{Q})\hat{R}$ is the QR decomposition of $\tilde{S}$. In other words, the R-factor of $R + F$ yields the R-factor for $\tilde{S}$.

Let P_1 be the Householder matrix used in the first step of the QR factorization of $R + F$, i.e.

$$
P_1 = I - \beta_1 \mathbf{w_1} \mathbf{w_1^T},
$$

where

$$
\beta_1 = \frac{2}{\| \mathbf{w_1} \|_2^2}
$$

and

$$
\mathbf{w_1} = r_{1,1}\mathbf{e_1} + F\mathbf{e_1} \pm \alpha_1 \mathbf{e_1},
$$

$r_{1,1}$ is the upper left entry of R and α_1 is the 2-norm of the first column of $R + F$. The plus/minus sign in front of α_1 is chosen so as to avoid cancellation. For the sake of simplicity, let us suppose $r_{1,1} > 0$; the correct choice is then $\mathbf{w_1} = r_{1,1}\mathbf{e_1} + F\mathbf{e_1} + \alpha_1 \mathbf{e_1}$.

Let $\mathbf{v_1} = 2r_{1,1}\mathbf{e_1}$. Notice that $\mathbf{v_1}$ is the Householder vector associated with the first column of R. The first fact we need to prove is that $\mathbf{w_1}$ is "close" to $\mathbf{v_1}$. We have:

$$
\begin{aligned}
\mathbf{w_1} - \mathbf{v_1} &= \mathbf{w_1} - 2r_{1,1}\mathbf{e_1} \\
&= F\mathbf{e_1} + \alpha_1 \mathbf{e_1} - r_{1,1}\mathbf{e_1} \\
&= \begin{bmatrix} f_{1,1} - r_{1,1} + \sqrt{r_{1,1}^2 + f_{1,1}^2 + 2r_{1,1}f_{1,1} + f_{2,1}^2 + \cdots + f_{N,1}^2} \\ f_{2,1} \\ \vdots \\ f_{N,1} \end{bmatrix}.
\end{aligned}
$$

Since

$$
\sum_i f_{i,1}^2 = \| F\mathbf{e_1} \|_2^2 \leq \| F \|_2^2 = \| E \|_2^2 \leq \eta^2,
$$

then

$$\left|\sqrt{r_{1,1}^2 + f_{1,1}^2 + 2r_{1,1}f_{1,1} + f_{2,1}^2 + \cdots + f_{N,1}^2} - r_{1,1}\right|$$
$$\leq \left|\sqrt{r_{1,1}^2 + 2r_{1,1}f_{1,1} + \eta^2} - r_{1,1}\right|$$
$$\leq \left|\sqrt{r_{1,1}^2 + 2r_{1,1}\eta + \eta^2} - r_{1,1}\right|$$
$$\leq \left|\sqrt{(r_{1,1} + \eta)^2} - r_{1,1}\right|$$
$$= \eta.$$

In this way we have

$$\|\mathbf{w}_1 - \mathbf{v}_1\|_2 \leq \|F\mathbf{e}_1\|_2 + \eta \leq 2\eta.$$

Set

$$\mathbf{w}_1 = \mathbf{v}_1 + \tilde{\mathbf{v}}_1, \qquad (6.3.1)$$

where $\|\tilde{\mathbf{v}}_1\|_2 \leq 2\eta$.

The first Householder triangularization step on $R + F$ is defined as

$$P_1(R + F) = (I - \beta_1 \mathbf{w}_1 \mathbf{w}_1^T)R + P_1 F.$$

Since P_1 is orthogonal, $P_1 F$ has small norm:

$$\|P_1 F\|_2 = \|F\|_2 \leq \eta.$$

We will perform a first-order approximation on the term $(I - \beta_1 \mathbf{w}_1 \mathbf{w}_1^T)R$. Using (6.3.1), we have:

$$(I - \beta_1 \mathbf{w}_1 \mathbf{w}_1^T)R = [I - \beta_1(\mathbf{v}_1 + \tilde{\mathbf{v}}_1)(\mathbf{v}_1^T + \tilde{\mathbf{v}}_1^T)]R$$
$$\doteq (I - \beta_1 \mathbf{v}_1 \mathbf{v}_1^T)R - \beta_1(\mathbf{v}_1 \tilde{\mathbf{v}}_1^T - \tilde{\mathbf{v}}_1 v_1^T)R$$
$$= R_1 + \beta_1 \tilde{\mathbf{v}}_1 \mathbf{v}_1^T R,$$

where $R_1 = (I - \beta_1 \mathbf{v}_1 \mathbf{v}_1^T - \beta_1 \mathbf{v}_1 \tilde{\mathbf{v}}_1^T)R$. Notice that R_1 is triangular, because $I - \beta_1 \mathbf{v}_1 \mathbf{v}_1^T$ is a diagonal matrix, of the form

$$I - \beta_1 \mathbf{v}_1 \mathbf{v}_1^T = \begin{bmatrix} 1 - \dfrac{8r_{1,1}^2}{\|\mathbf{w}_1\|_2^2} & & & \\ & 1 & & \\ & & \ddots & \\ & & & 1 \end{bmatrix}, \qquad (6.3.2)$$

and $\mathbf{v}_1 \tilde{\mathbf{v}}_1^T$ is the zero matrix except for the first row.

Due to the form of the matrix $\tilde{\mathbf{v}}_1\mathbf{v}_1^T$, which is zero except for the first column, the norm of the term $\beta_1\tilde{\mathbf{v}}_1\mathbf{v}_1^T R$ can be written as:

$$
\begin{aligned}
\|\beta_1\tilde{\mathbf{v}}_1\mathbf{v}_1^T R\|_2 &\leq 2\beta_1\eta\|\mathbf{v}_1\|_2\|R_{(1,:)}\|_2 \\
&= \frac{4}{\|\mathbf{w}_1\|_2^2}\eta\|\mathbf{v}_1\|_2\|R_{(1,:)}\|_2 \\
&= \frac{4}{\|\mathbf{w}_1\|_2^2}\eta\|\mathbf{w}_1 - \tilde{\mathbf{v}}_1\|_2\|R_{(1,:)}\|_2 \\
&\leq \frac{4}{\|\mathbf{w}_1\|_2^2}\eta(\|\mathbf{w}_1\|_2 + \|\tilde{\mathbf{v}}_1\|_2)\|R_{(1,:)}\|_2 \\
&\doteq \frac{4}{\|\mathbf{w}_1\|_2^2}\eta\|\mathbf{w}_1\|_2\|R_{(1,:)}\|_2 \\
&= \frac{4}{\|\mathbf{w}_1\|_2}\eta\|R_{(1,:)}\|_2,
\end{aligned}
$$

where a Matlab-like notation $R_{(1,:)}$ has been used.

It is not possible to give an a priori bound on $\|\mathbf{v}_1\|_2$ or $\|\mathbf{w}_1\|_2$. But it follows from the above calculation that

$$
\|\beta_1\tilde{\mathbf{v}}_1\mathbf{v}_1^T R\|_2 \doteq \frac{4}{|\hat{R}_{(1,1)}|}\eta\|\hat{R}_{(1,:)}\|_2
$$

and $\hat{r}_{1,1}$ and $\hat{R}_{(1,:)}$ are known from the explicit computation of the R factor for $\tilde{S}$.

As for the term $P_1 F$, we have already observed that it has small norm; also notice that, in a first order approximation, we have

$$
P_1 F \doteq (I - \beta_1\mathbf{v}_1\mathbf{v}_1^T)F. \tag{6.3.3}
$$

So far we have proved that

$$
P_1(R + F) = R_1 + F_1,
$$

where R_1 is upper triangular and $F_1 = \beta_1\tilde{\mathbf{v}}_1\mathbf{v}_1^T R + P_1 F$ is such that

$$
\|F_1\|_2 \leq \eta + \frac{4}{|\hat{R}_{(1,1)}|}\eta\|\hat{R}_{(1,:)}\|_2.
$$

The above argument can be applied to the subsequent steps of the House-holder QR algorithm. In the second step, the Householder matrix P_2 acts on the $(N - 1) \times (N - 1)$ right lower submatrix of $R_1 + F_1$. We will denote with $\bar{M}$ the $(N-1) \times (N-1)$ right lower submatrix of an $N \times N$ matrix M. It follows from (6.3.2) and (6.3.3) that

$$\overline{(R_1 + F_1)} = \bar{R}_1 + \bar{F}_1$$
$$= \bar{R} + \bar{F}_1,$$
$$\text{where} \quad \bar{F}_1 = \bar{F} + \overline{\beta \tilde{v}_1 v_1^T R}.$$

This means that, if $\frac{\|\hat{R}_{1,:}\|_2}{|\hat{r}_{1,1}|}$ is not too large, the second Householder step acts on the sum of a matrix of small norm, plus a triangular matrix which is a submatrix of the same R we started from. Now, the second Householder matrix is

$$P_2 = I - \beta_2 \mathbf{w}_2 \mathbf{w}_2^T$$
$$= I - \beta_2 (\mathbf{v}_2 + \tilde{\mathbf{v}}_2)(\mathbf{v}_2^T + \tilde{\mathbf{v}}_2^T),$$

where $\mathbf{w}_2$, $\mathbf{v}_2$ and $\tilde{\mathbf{v}}_2$ are defined as in step 1.

Under the hypothesis that $\frac{\|\hat{R}_{1,:}\|_2}{|\hat{r}_{1,1}|}$ is not large, $\tilde{\mathbf{v}}_2$ has small norm, since we have $\|\tilde{\mathbf{v}}_2\|_2 \leq \eta + \frac{4}{|\hat{r}_{1,1}|}\eta\|\hat{R}_{1,:}\|_2$.

So the second step can be written as:

$$P_2(\bar{R} + \bar{F} + \overline{\beta \tilde{v}_1 v_1^T R}) = R_2 + F_2$$

with

$$R_2 = (I - \beta_2 \mathbf{v}_2 \mathbf{v}_2^T - \beta_2 \tilde{\mathbf{v}}_2 \mathbf{v}_2^T)\bar{R}_1$$
$$F_2 = -\beta_2 \mathbf{v}_2 \tilde{\mathbf{v}}_2^T \bar{R}_1 + P_2 \overline{\beta_1 \tilde{v}_1 v_1^T)R} + P_2 \bar{F},$$

where:

- R_2 is upper triangular and $\bar{R}_2$ is the $(N-2) \times (N-2)$ right lower submatrix of R;
- F_2 has small norm, provided that $\frac{\|\hat{R}_{(2,:)}\|_2}{|\hat{R}_{(2,2)}|}$ is not large. Indeed we have:

$$\|F_2\|_2 \lesssim \|\beta_2 \mathbf{v}_2 \tilde{\mathbf{v}}_2^T \bar{R}_1\|_2 + \|(I - \beta_2 \mathbf{v}_2 \mathbf{v}_2^T)\overline{\beta_1 \tilde{v}_1 v_1^T)R}\|_2 + \|F\|_2$$

and therefore

$$\|F_2\|_2 \lesssim \left(4\frac{\|\hat{R}_{(2,:)}\|_2}{|\hat{R}_{(2,2)}|} + 1\right)\left(4\eta\frac{\|\hat{R}_{(1,:)}\|_2}{|\hat{R}_{(1,1)}|} + \eta\right).$$

In general, the k-th Householder step is given by

$$P_k(\bar{R}_{k-1} + \bar{F}_{k-1}) = R_k + F_k,$$

where R_k is upper triangular and coincides, except for its first row, with a right lower submatrix of R, and F_k is such that

$$\|F_k\|_2 \leq \eta \prod_{i=1}^{k-1} \left(1 + 4\frac{|R_{(i,i+1)}|}{|R_{(i,i)}|}\right) \left(1 + 4\frac{\|R_{(k,:)}\|_2}{|R_{(k,k)}|}\right). \tag{6.3.4}$$

Stability is ensured if the quantities

$$\xi_k = \frac{\|\hat{R}_{(k,:)}\|_2}{|\hat{R}_{(k,k)}|},$$

which will be called *amplification factors*, remain small throughout the factorization, or at least until the last row of significant magnitude in R is computed.

The formula that gives amplification factors also suggests that a diagonal entry $\hat{r}_{i,i}$ of small absolute value occurring in an early decomposition stage may compromise stability. This can be avoided by performing QR *decomposition with pivoting*: compute $S = QR\Pi$, where Π is a permutation matrix and the triangular factor R has diagonal entries of decreasing absolute value.

Notice that formula (6.3.4) has been derived for Sylvester matrices, but only a minor adjustment is needed so that it can be applied to the Bezoutian. The above discussion about amplification factors and the use of pivoting is therefore valid for the Bezoutian as well.

6.3.2. Degree of an ϵ-GCD

We next seek to recover the degree of an ϵ-GCD from the computed R. As will be explained later, it suffices to find an upper bound on the degree.

Formula (6.3.4) does yield such a bound. Indeed, let k_0 be the minimum value of k such that (6.3.4) holds with F_k replaced by the trailing submatrix of R of size $(N - k) \times (N - k)$, where $N = m + n$ or $N = n$ depending on whether the Sylvester or Bézout matrix is being factorized. Then the ϵ-GCD degree is less than or equal to $N - k_0$.

Unfortunately, this estimate usually turns out to be too pessimistic for practical application. One may try to use global estimates on the matrices involved in the QR decomposition, rather than analyzing the decomposition step by step. With the same notation as before, let

$$A = R + F = \hat{Q}\hat{R}, \tag{6.3.5}$$

where A is an $N \times N$ resultant matrix and R is singular. Let k be the nullity of R (and the degree of an approximate divisor) and rewrite (6.3.5) dividing the matrices into blocks of sizes

$$\begin{array}{c|c} (N-k) \times (N-k) & (N-k) \times k \\ \hline k \times (N-k) & k \times k \end{array}.$$

We obtain

$$\begin{array}{c|c} A_1 & A_2 \\ \hline A_4 & A_3 \end{array} = \begin{array}{c|c} \hat{Q}_1 & \hat{Q}_2 \\ \hline \hat{Q}_4 & \hat{Q}_3 \end{array} \begin{array}{c|c} \hat{R}_1 & \hat{R}_2 \\ \hline 0 & \hat{R}_3 \end{array}, \tag{6.3.6}$$

where $\|A_3\| \le \eta$ and $\|A_4\| \le \eta$. We give an estimate on the norm of $\hat{R}_3$:

Proposition 6.3.2. *With the above notation, the following estimate holds at first order in η:*

$$\|\hat{R}_3\|_2 \le \eta(1 + \|\hat{R}_1^{-1}\|_2 \|\hat{R}_2\|_2). \tag{6.3.7}$$

Proof. We have

$$\hat{Q}_4 \hat{R}_1 = A_4,$$

hence

$$\|\hat{Q}_4\|_2 \le \|\hat{R}_1^{-1}\|_2 \|A_4\|_2 \le \eta \|\hat{R}_1^{-1}\|_2. \tag{6.3.8}$$

The block decomposition also yields

$$A_3 = \hat{Q}_4 \hat{R}_2 + \hat{Q}_3 \hat{R}_3,$$

and therefore

$$\hat{R}_3 = \hat{Q}_3^{-1}(A_3 - \hat{Q}_4 \hat{R}_2). \tag{6.3.9}$$

By taking norms in (6.3.9) and substituting (6.3.8) we obtain:

$$\|\hat{R}_3\|_2 \le \eta \|\hat{Q}_3^{-1}\|_2 (1 + \|\hat{R}_1^{-1}\|_2 \|\hat{R}_2\|_2).$$

Now, $\hat{Q}_3$ is close to being orthogonal, because $\hat{Q}_4$ has small norm. More precisely, from the block decomposition and from the orthogonality of $\hat{Q}$ it follows that:

$$\hat{Q}_4 \hat{Q}_4^T + \hat{Q}_3 \hat{Q}_3^T = I$$
$$\hat{Q}_3 \hat{Q}_3^T = I - \hat{Q}_4 \hat{Q}_4^T$$
$$\hat{Q}_3^T = \hat{Q}_3^{-1}(I - \hat{Q}_4 \hat{Q}_4^T)$$
$$\hat{Q}_3^{-1} = \hat{Q}_3^T(I - \hat{Q}_4 \hat{Q}_4^T)-1$$
$$\|\hat{Q}_3^{-1}\|_2 \le \|\hat{Q}_3^T\|_2 \|(I - \hat{Q}_4 \hat{Q}_4^T)-1\|_2$$

Notice that $\hat{Q}_3^T$ is a submatrix of the orthogonal matrix $\hat{Q}^T$, therefore its norm is less or equal to 1. As for $(I - \hat{Q}_4\hat{Q}_4^T)$, we have $\|(I - \hat{Q}_4\hat{Q}_4^T) - 1\|_2 \dot{<} 1 + \|\hat{Q}_4\|_2\|\hat{Q}_4^T\|_2$; therefore we have $\|\hat{Q}_3^{-1}\|_2 \dot{=} 1$ at first order in η. Hence the thesis. $\qquad\square$

This result says that whenever the block sizes in (6.3.6) are chosen so that (6.3.7) is not satisfied, then $u(x)$ and $v(x)$ have no ϵ-divisor of degree less or equal to the size of $\hat{R}_3$. Therefore the largest size of R_3 such that (6.3.7) holds is also an upper bound for the GCD degree.

If necessary, (6.3.7) can be rewritten using 1- or ∞-norms, which are easier to compute than the 2-norm.

An easier –but heuristic– version of the bound given in (6.3.7) can be obtained observing that, as a consequence of pivoting in the QR decomposition, the diagonal entries $R_{k,k}$ have moduli that decrease with k and, moreover, $|R_{k,k}|$ dominates, for every k, the moduli of the elements in the k-th row. As a consequence, the quantity $1/|R_{k,k}|$ can be chosen as an estimator for $\|\hat{R}_1^{-1}|_2$. Notice that, with this choice, the resulting bound might, in principle, be no longer an upper bound. We have to use a heuristic approach here, because a more precise statement would necessary involve an exponential factor. So, (6.3.7) becomes

$$\|\hat{R}_3\|_2 \leq \eta \left(1 + \frac{\|\hat{R}_2\|_2}{|R_{(k,k)}|} \right). \qquad (6.3.10)$$

This heuristic bound has been used in numerical tests in Chapter 8 and has proven to be quite effective.

6.3.3. Coefficients of an ϵ-GCD

The drawback of QR decomposition with pivoting is that it no longer automatically gives a set of coefficients for the GCD. So, once the degree is known, other techniques must be employed to find a set of coefficients.

Suppose that

$$\begin{cases} u(x) = g(x)p(x) \\ v(x) = g(x)q(x) \end{cases} \qquad (6.3.11)$$

It follows that

$$u(x)q(x) - v(x)p(x) = 0. \qquad (6.3.12)$$

Then (6.3.12) can be written in matrix form as:

$$
\begin{pmatrix}
u_0 & & & & v_0 & & & \\
u_1 & \ddots & & & v_1 & \ddots & & \\
\vdots & & u_0 & \vdots & & & v_0 & \\
\vdots & & u_1 & \vdots & & & v_1 & \\
u_n & & & \vdots & v_m & & & \vdots \\
& \ddots & & \vdots & & \ddots & & \vdots \\
& & u_n & & & & v_m &
\end{pmatrix}
\begin{pmatrix}
q_0 \\
q_1 \\
\vdots \\
q_{m-k} \\
-p_0 \\
-p_1 \\
\vdots \\
-p_{n-k}
\end{pmatrix}
= 0. \qquad (6.3.13)
$$

A solution for this linear system gives the cofactors $p(x)$ and $q(x)$. Then $g(x)$ can be retrieved as either $g(x) = u(x)/p(x)$ or $g(x) = v(x)/p(x)$.

Since (6.3.13) is a homogeneous system, normalization is needed. In practical implementations we have chosen a monic $p(x)$

Remark 6.3.3. The customary polynomial division algorithm is known to be numerically unstable. It is therefore advisable to perform polynomial divisions either using FFT (*i.e.* with evaluation/interpolation techniques, see [19]), or by solving a least squares problem; see Sections B.1.1 and B.2 for details.

6.3.4. Iterative refinement

As above, suppose that $g(x)$ is a common factor of the polynomials $u(x)$ and $v(x)$, so that (6.3.11) holds.

For any polynomial $f(x)$, denote by $\mathbf{f}$ the vector $[f_0 \ f_1 \ \cdots \ f_{\deg f}]^T$ whose entries are the coefficients of $f(x)$. With this notation, the system (6.3.11) can be written as

$$
F(\mathbf{z}) = A(\mathbf{z}) - \mathbf{w} = 0, \qquad (6.3.14)
$$

where

$$
\mathbf{z} = \begin{bmatrix} \mathbf{g} \\ \mathbf{p} \\ \mathbf{q} \end{bmatrix}, \ \mathbf{w} = \begin{bmatrix} \mathbf{u} \\ \mathbf{v} \end{bmatrix},
$$

$$
A(\mathbf{z}) = \begin{bmatrix} \mathcal{C}(p)\mathbf{g} \\ \mathcal{C}(q)\mathbf{g} \end{bmatrix} = \begin{bmatrix} \mathcal{C}(g)\mathbf{p} \\ \mathcal{C}(g)\mathbf{q} \end{bmatrix}
$$

and $\mathcal{C}(p), \mathcal{C}(q)$ and $\mathcal{C}(g)$ are convolution matrices of appropriate size.

When looking for an approximate GCD, however, the right-hand side of (6.3.14) is expected to be nonzero; we should solve a least squares system where the quantity

$$F(\mathbf{z}) = A(\mathbf{z}) - \mathbf{w} = \mathbf{r} \tag{6.3.15}$$

is to be minimized.

This can of course be done by a standard optimization method (such as, for instance, the Levenberg-Marquardt method). However, we choose here to use Newton's method(also known as the Gauss-Newton method in the optimization literature).

Each iterative step can be written as

$$\mathbf{z}_{j+1} = \mathbf{z}_j - \eta_j, \tag{6.3.16}$$

where η_j solves the linear least-squares problem

$$J(\mathbf{z}_j)\mathbf{x} = A\mathbf{z}_j - \mathbf{w}, \tag{6.3.17}$$

that is, we have[1]

$$\eta_I = J^{\dagger}(A\mathbf{z}_j - \mathbf{w}), \tag{6.3.18}$$

where $J^{\dagger}$ is the Moore-Penrose pseudoinverse of the matrix J.

Notice that the Jacobian matrix J associated with A can be easily computed as

$$J = \begin{pmatrix} \mathcal{C}(p) & \mathcal{C}(g) & 0 \\ \mathcal{C}(q) & 0 & \mathcal{C}(g) \end{pmatrix}, \tag{6.3.19}$$

where $\mathcal{C}(p)$ is $n \times (k+1)$, $\mathcal{C}(q)$ is $m \times (k+1)$, $\mathcal{C}(g)$ in the first block row is $n \times (n - k + 1)$ and $\mathcal{C}(g)$ in the second block row is $m \times (m - k + 1)$.

There is, however, a difficulty that will now be discussed.

Remark 6.3.4. Under the hypotheses stated above, the Jacobian matrix (6.3.19) computed in any point $\mathbf{z} = [\mathbf{g} \quad \mathbf{p} \quad \mathbf{q}]^T$ is singular. Moreover, the nullity (that is, the dimension of the null space) of J is 1 if and only if $p(x)$, $q(x)$ and $g(x)$ have no common factors. In particular, if $\mathbf{z}$ is a solution of the system (6.3.14) and $g(x)$ has maximum degree, *i.e.*, it is a GCD, then J has nullity one and any vector in the null space of J is a multiple of $[\mathbf{g} \quad -\mathbf{p} \quad -\mathbf{q}]^T$, where $p(x)$ and $q(x)$ are cofactors.

---

[1] As a motivation for (6.3.18), we recall here that, given a smooth nonlinear function $\mathcal{F} : \mathbb{R}^M \to \mathbb{R}^N$, with $M > N$, minimizing $\mathcal{G}(\mathbf{z}) = \|\mathcal{F}(\mathbf{z})\|^2$ requires to compute critical points $\mathbf{z}_c$ such that $J_{\mathcal{G}}(\mathbf{z}_c) = 0$, where $J_{\mathcal{G}}(\mathbf{z})$ is the Jacobian of $\mathcal{G}(\mathbf{z})$. This is equivalent to solving the generalized normal equations $J_{\mathcal{F}}(\mathbf{z})^* \mathcal{F}(\mathbf{z}) = 0$, which in turn, by using elementary properties of the pseudoinverse, can be shown to be equivalent to $J_{\mathcal{F}}(\mathbf{z})^{\dagger} \mathcal{F}(\mathbf{z}) = 0$. Under proper hypotheses (*e.g.* analyticity), the same holds in the complex case.

Proof. The Jacobian matrix J is singular if and only if there exists a nonzero vector $\mathbf{t}$ such that $J\mathbf{t} = \mathbf{0}$. It follows from (6.3.19) that this is equivalent to the existence of polynomials $a(x)$, $b(x)$ and $c(x)$ of degrees respectively k, $n - k$ and $m - k$, such that

$$p(x)a(x) + g(x)b(x) = 0,$$
$$q(x)a(x) + g(x)c(x) = 0.$$

The choice $a(x) = g(x)$, $b(x) = -p(x)$ and $c(x) = -q(x)$ clearly satisfies the above equations.

As for the second statement, assume that $g(x)$ has no common factor with both $p(x)$ and $q(x)$. Then $g(x)$ divides $a(x)$; but since they have the same degree, they are equal up to multiplication by a constant. Therefore $b(x)$ and $c(x)$ are also uniquely determined. It follows that, up to scaling, there is only one vector in the null space of J.

Conversely, let $s(x)$ be a polynomial of positive degree such that $p(x) = p_1(x)s(x)$, $q(x) = q_1(x)s(x)$ and $g(x) = g_1(x)s(x)$. Of course the choice $a(x) = g(x)$, $b(x) = -p(x)$, $c(x) = -q(x)$ still gives a solution, but we might also set $a(x) = g_1(x)\zeta(x)$, $b(x) = p_1(x)\zeta(x)$ and $c(x) = q_1(x)\zeta(x)$, where $\zeta(x)$ is an arbitrary polynomial of the same degree as $s(x)$. Therefore there are at least two linearly independent vectors in the null space of J.

Finally, observe that $g(x)$ is a GCD iff $p(x)$ and $q(x)$ have no common factors. This implies that if $g(x)$ is a GCD, then J has nullity one.   □

In general, the singularity of the Jacobian matrix may pose difficulties to the application of Newton's method, such as the non-uniqueness of the minimum and a slow convergence rate (linear instead of quadratic in the zero-residual case).

The non-uniqueness of the minimum is related to the fact that the system (6.3.14) defines a GCD only up to a scaling factor. In other words, if $\mathbf{z} = [\mathbf{g} \quad \mathbf{p} \quad \mathbf{q}]^T$ is a solution of (6.3.14), then $\mathbf{z}_\phi = [\phi\mathbf{g} \quad \phi^{-1}\mathbf{p} \quad \phi^{-1}\mathbf{q}]^T$ is also a solution for any complex number ϕ. (A detailed geometric description of this situation can be found in [50]).

A way to avoid the presence of this free parameter is to add an equation to (6.3.14), so as to impose some kind of normalization to the GCD (as is done in [144]). For example, one might require $g(x)$ to be monic, or to have unitary norm. If the normalization is properly chosen (that is, if the coefficients in the additional equation form a vector that is not orthogonal to $\mathbf{g}$), then the Jacobian turns out to be nonsingular. Choosing a normalization can be tricky, because each choice might work well for some problems but not for others. For instance, requiring $g(x)$ to be monic might not be a good idea if $g(x)$ has highly unbalanced coefficients. Besides, if the systems (6.3.14) and (6.3.15) are modified, then the residual $\mathbf{r}$ in (6.3.15) is no longer a valid parameter to determine whether an ϵ-GCD is actually being computed, according to the definition given in Section 1.1.2; so one must remember to compute at each step the actual perturbation of the input polynomials given by the iterative method.

Alternatively, one may choose a normalization constraint and incorporate it into (6.3.14) and (6.3.15) before proceeding to iterative refinement. This technique has the advantage of reducing the number of equations and unknowns in the least squares problem (6.3.15). However, one should choose a normalization given by a very simple constraint on the coefficients of the GCD or cofactors (e.g. requiring $g(x)$ to be monic), otherwise the required computations will grow complicated and expensive. Such simple constraints usually are not the best choice in terms of numerical stability.

For these reasons, in order to achieve better stability and convergence properties, we force the Jacobian to have full rank by adding a row, given by $\mathbf{w}^T = [\mathbf{g}^T \quad -\mathbf{p}^T \quad -\mathbf{q}^T]$. Nevertheless, it can be proved, by relying on the results of [110], that the quadratic convergence of Newton's method in the case of zero residual also holds, in this case, with a rank deficient Jacobian. This property, which is discussed in Section 6.3.5, is useful when the initial guess for the approximate GCD degree is too small, since in this case the rank deficiency of the Jacobian is unavoidable. This is not the case in the algorithms described in this chapter, since they rely on upper bounds on the GCD degree, but in the algorithm Fastgcd (see next chapter) the initial guess on the degree might conceivably be too small.

The new Jacobian $\tilde{J} = \left(\begin{smallmatrix} J \\ \mathbf{w}^T \end{smallmatrix}\right)$ is associated with the least squares problem that minimizes $\tilde{F}(\mathbf{z}) = \left(\begin{smallmatrix} F(\mathbf{z}) \\ \|\mathbf{g}\|^2 - \|\mathbf{p}\|^2 - \|\mathbf{q}\|^2 - K \end{smallmatrix}\right)$, where K is a constant. The choice of $\mathbf{w}^T$ as an additional row helps to ensure that the solution of each Newton step

$$\mathbf{z}_{j+1} = \mathbf{z}_j - \tilde{J}(\mathbf{z}_j)^\dagger \tilde{F}(\mathbf{z}_j) \qquad (6.3.20)$$

is nearly orthogonal to ker J. For ease of notation, the new Jacobian will be denoted simply as J in the following.

The described method can be used to refine the results obtained in the previous sections. Suppose that a tentative divisor $g_0(x)$ with associated vector $\mathbf{g}_0$ has been computed, together with cofactors $p_0(x)$ and $q_0(x)$. Use the vector

$$\mathbf{z}_0 = \begin{bmatrix} \mathbf{g}_0 \\ \mathbf{p}_0 \\ \mathbf{q}_0 \end{bmatrix}$$

as starting point for Newton's method. At each iteration step, the "near-ness" $\delta = [\|\hat{u}(x) - p(x)g(x)\|_2, \|\hat{v} - q(x)g(x)\|_2]$ is to be computed. The iteration stops when $\|\delta\|_2$ either stops decreasing[2], or becomes very small (comparable to the machine precision, say), or if a fixed maximum number of iterations has been reached. The result provides a new divisor and cofactors, which minimize the residual associated with the system (6.3.14).

If both entries of δ are smaller than ϵ, an ϵ-divisor of degree k has been found; otherwise we must conclude that either no such divisor exists, or the method has failed. Failure may be caused by insufficient accuracy of the initial approximation, and/or by the lack of global convergence of the iterative refinement procedure, which might output a local (nonglobal) minimum.

There are two main reasons why iterative refinement is important for our method. The first one is, of course, that iterative refinement allows for a remarkable improvement in the accuracy of the results. The second reason is that, since δ is computed at each step, if both entries of δ are found to be smaller than ϵ then $g(x)$ is certified as an ϵ-divisor of $\hat{u}(x)$ and $\hat{v}(x)$.

Saying that $g(x)$ is an ϵ-divisor of $\hat{u}(x)$ and $\hat{v}(x)$ does not mean that it is an ϵ-GCD, because it might not have maximum degree. As shown in Section 6.3.2 , though, an upper bound k_0 on the degree of an ϵ-GCD can be easily computed. One may then proceed as follows:

- construct and solve the system (6.5.5) so that a polynomial of degree $k = k_0$ is obtained;
- perform iterative refinement;
- if iterative refinement outputs an ϵ-divisor, then that is also an ϵ-GCD, and the algorithm stops. Otherwise, decrease k by 1 and repeat the procedure until either an ϵ-GCD is found, or $k = 0$, in which case $\hat{u}(x)$ and $\hat{v}(x)$ are ϵ-coprime.

6.3.5. Rank-deficient Jacobian

It follows from Remark 6.3.4 that the Jacobian matrix is singular not only when computed in the solution of (6.3.15), but also with respect to each iterate z_j. An analysis of Newton's method applied to least squares problems of this type - called *uniformly rank deficient* problems - can

---

[2] In principle, we cannot always expect a monotonous behaviour from δ. Likely, the stopping criterion we are using here could be replaced by a more effective one: for instance, one might include a line-search strategy (see Section 7.4.1).

be found in [51]. At each iteration (6.3.16), one has to solve the linear least-squares problem (6.3.17), whose solution is not unique; therefore one must decide how to choose the component of $\mathbf{x}$ that lies in the null space of $J^{\dagger}(\mathbf{z}_j)$. The authors of [51] seek a minimum norm solution of the nonlinear problem (or, more generally, a solution that minimizes the distance from a predefined center vector $\mathbf{c}$) and choose the solution of (6.3.17) accordingly.

However, we have no particular reason to seek a minimum norm solution of (6.3.15); besides, the choice of $\mathbf{c}$ would raise the same difficulties described above when discussing the balancing of the GCD. We are rather interested in improving the convergence properties of the iteration (6.3.16). The most natural choice (see also [53]) is then to always compute the minimum norm solution of (6.3.17), which is equivalent to writing (6.3.16) in the well-known general form

$$\mathbf{z}_{j+1} = \mathbf{z}_j - J^{\dagger}(\mathbf{z}_j)(A\mathbf{z}_j - \mathbf{w}), \qquad (6.3.21)$$

where $J^{\dagger}(\mathbf{z}_j)$ is the Moore-Penrose pseudoinverse of $J(\mathbf{z}_j)$. Let us now examine the convergence of (6.3.21).

In the zero-residual case, we claim that the convergence of (6.3.21) is quadratic, even though the Jacobian is singular.

In the nonsingular case, the quadratic convergence of Newton's method can be proved (see *e.g.* [54]) by estimating the distance between the descent direction provided by Newton's method and the direction provided by standard multidimensional Newton's method (which is known to have quadratic convergence). In the singular case, a straightforward application of the same approach fails because the convergence of the Newton process is linear. However, Rall ([110]) investigated the behaviour of the multidimensional Newton process when the Jacobian computed in the solution is singular – a case where the convergence is known to be linear – and formulated a correction of the method that restores the quadratic convergence[3].

Consider a Newton iteration

$$\mathbf{z}_{n+1} = \mathbf{z}_n - J(\mathbf{z}_n)^{\dagger} P(\mathbf{z}_n). \qquad (6.3.22)$$

---

[3] Rall actually examines the iteration

$$\mathbf{z}_{n+1} = \mathbf{z}_n - J(\mathbf{z}_n)^{-1} P(\mathbf{z}_n),$$

where the function $P : X \longrightarrow X$ (with $X = \mathbb{R}^N$ or $\mathbb{C}^N$) is such that $J(\mathbf{z}_n)$ is nonsingular for every n (so that $J(\mathbf{z}_n)^{-1}$ is well-defined). We aim here to adapt his proof to the case of a rectangular system with rank-deficient Jacobians.

Let $\mathbf{z}^*$ be the limit point of the iteration, J the Jacobian matrix computed in $\mathbf{z}^*$. As stated in Remark 6.3.4, in the case we are interested in, J is rank-deficient and has nullity one.[4] Also, the second derivative operator is nonsingular: indeed, in Rall's notation, we have

$$P'' \begin{pmatrix} \alpha \\ \beta \\ \gamma \end{pmatrix} = \begin{pmatrix} C_\beta & C_\alpha & 0 \\ C_\gamma & 0 & C_\alpha \end{pmatrix} \neq 0,$$

where the vectors α, β and γ are assumed to have the same lengths as $\mathbf{g}$, $\mathbf{p}$ and $\mathbf{q}$, respectively.

Under these hypotheses, the adaptation of Rall's result gives:

Proposition 6.3.5. *Let $\mathcal{N} \subset \mathbb{C}^M$ be the null space of J and $\mathcal{X}$ the maximal subspace of $\mathbb{C}^M$ orthogonal to $\mathcal{N}$, so that $\mathbb{C}^M = \mathcal{X} \oplus \mathcal{N}$. Assume that the corrected Newton process*

$$\tilde{\mathbf{z}}_{n+1} = \tilde{\mathbf{z}}_n - [\Pi_\mathcal{X}(J(\tilde{\mathbf{z}}_n)^\dagger P(\tilde{\mathbf{z}}_n))] - 2[\Pi_\mathcal{N}(J(\tilde{\mathbf{z}}_n)^\dagger P(\tilde{\mathbf{z}}_n))] \quad (6.3.23)$$

converges, where $\Pi_\mathcal{X}$ and $\Pi_\mathcal{N}$ are the projection operators from $\mathbb{C}^M$ on $\mathcal{X}$ and $\mathcal{N}$, respectively, and the pseudoinverses $J(\tilde{\mathbf{z}}_n)^\dagger$ are assumed to be uniformly bounded. Then the convergence is quadratic.

Let us describe the situation in detail. For every n, let $\epsilon_n = \mathbf{z}_n - \mathbf{z}^*$ and $\eta_n = J(\mathbf{z}_n)^\dagger P(\mathbf{z}_n)$. Write

$$\epsilon_n = \xi_n + \zeta_n \quad \text{with} \quad \xi_n \in \mathcal{X}, \quad \zeta_n \in \mathcal{N}. \quad (6.3.24)$$

Then Rall proves that

$$\xi_{n+1} = \mathcal{O}(\|\epsilon_n\|_2^2) \quad (6.3.25)$$

whereas

$$\zeta_{n+1} = \frac{1}{2}\zeta_n + \delta_{n+1} \quad (6.3.26)$$

where $\delta_{n+1} = \mathcal{O}(\|\epsilon_n\|_2^2)$. In other words, the component of ϵ_n that lies in $\mathcal{X}$ converges quadratically to zero, whereas the component that lies in $\mathcal{N}$ converges linearly, with rate $1/2$. But ζ_n can be made to converge quadratically if a correction is applied to the iteration, exactly as it happens in one-dimensional Newton's method. Indeed, in the non-corrected

---

[4] Not only J, but also all the $J(\mathbf{z}_n)$'s have nullity one. This is a necessary condition for the continuity of the pseudoinverse of the Jacobian. Moreover, we may assume $\|J(\mathbf{z})^\dagger\| = \|J\| + \mathcal{O}(\|\epsilon_n\|)$ for n large enough (see *e.g.* [122]).

case we have:

$$
\begin{aligned}
\mathbf{z}_n - \mathbf{z}^* = \eta_n &= \epsilon_{n+1} - \epsilon_n \\
&= \xi_{n+1} + \zeta_{n+1} - \xi_n - \zeta_n \\
&= -\frac{1}{2}\zeta_n - \xi_n + \delta, \qquad \text{with} \quad \|\delta\| = \mathcal{O}(\|\epsilon_n\|_2^2)
\end{aligned}
$$

but if (6.3.23) is applied, we obtain

$$
\begin{aligned}
\tilde{\epsilon}_{n+1} &= \tilde{\epsilon}_n - \Pi_{\mathcal{X}}(\eta_n) - 2\Pi_{\mathcal{N}}(\eta_n) \\
&= \xi_n + \zeta_n - \xi_n - 2 \cdot \frac{1}{2}\zeta_n - \Pi_{\mathcal{X}}(\delta) - 2\Pi_{\mathcal{N}}(\delta)
\end{aligned}
$$

and therefore

$$
\|\tilde{\epsilon}_{n+1}\|_2 = \mathcal{O}(\|\tilde{\epsilon}_n\|_2^2). \tag{6.3.27}
$$

The main difficulty with Rall's corrected method is that it requires to perform the projections $\Pi_{\mathcal{X}}$ and $\Pi_{\mathcal{N}}$ at each iteration, which usually means knowing the subspaces $\mathcal{X}$ and $\mathcal{N}$ beforehand. In our case, for example, if a vector in the null space of J were known before performing the iterative refinement, then Remark 6.3.4 implies that a GCD would already be available.

However, Rall's result is useful here because it helps to prove that the iteration (6.3.16) already converges quadratically, without any modification needed (see also Example 7.4.2). Indeed, if we wished to apply Rall's method to (6.3.16), the new Newton correction $\tilde{\eta}_n$ would be at each step

$$
\tilde{\eta}_n = \eta_n - \Pi_{\mathcal{N}}(\eta_n).
$$

Recall that $\eta_n = J(\mathbf{z}_n)^\dagger P(\mathbf{z}_n)$ is orthogonal to the null space of $J(\mathbf{z}_n)$, and in particular to the vector $\mathbf{t}_n = [\mathbf{g}_n \quad -\mathbf{p}_n \quad -\mathbf{q}_n]^T$. Let $\mathbf{t} = [\mathbf{g} \quad -\mathbf{p} \quad -\mathbf{q}]^T$ be a generator for $\mathcal{N}$. Moreover, let $\alpha_n = \mathbf{t} - \mathbf{t}_n$ and observe that $\|\alpha_n\|_2^2 = \|\mathbf{g} - \mathbf{g}_n\|_2^2 + \|\mathbf{p} - \mathbf{p}_n\|_2^2 + \|\mathbf{q} - \mathbf{q}_n\|_2^2 = \|\epsilon_n\|_2^2$. Then we have

$$
\begin{aligned}
\|\Pi_{\mathcal{N}}(\eta_n)\|_2 &= \frac{\eta_n^H \mathbf{t}}{\|\mathbf{t}\|_2^2} \\
&\leq \frac{1}{\|\mathbf{t}\|_2^2}\left(\eta_n^H(\alpha_n + \mathbf{t}_n)\right) \\
&= \frac{1}{\|\mathbf{t}\|_2^2}(\eta_n^H \alpha_n).
\end{aligned}
$$

Since $\|\eta_n\|_2 = \mathcal{O}(\|\epsilon_n\|_2)$, it follows that $\|\Pi_{\mathcal{N}}(\eta_n)\|_2 = \mathcal{O}(\|\epsilon_n\|_2^2)$. But then it must be $\|\epsilon_{n+1}\|_2 = \mathcal{O}(\|\epsilon_n\|_2^2)$.

This result agrees with the estimate obtained by considering the iteration (6.3.21) as a particular case of one of the solution techniques presented in [51] (namely the so-called truncated method, where the center vector is chosen at each iteration as the current iterate).

In the case when the residual is nonzero, the convergence of (6.3.21) is linear. Let H_i be the matrices of second derivatives of (6.3.14); then the convergence rate computed from [51] is given by the spectral radius of the matrix

$$K = (J^T J)^\dagger \sum z_i^* H_i, \tag{6.3.28}$$

which is proportional to the residual of the nonlinear least squares problem. Observe that (6.3.28) can be seen as a straightforward generalization of the well-known estimate for the convergence rate of ordinary Newton's method with nonzero residual.

Remark 6.3.6. The convergence properties of the Newton method examined above, together with experimental results, suggest that a small number of iterations should suffice to determine a local minimum. Unfortunately it is not possible to give a theoretical *a priori* bound on the number of iterations. Therefore, for a rigorous discussion of the GCD methods proposed in this chapter we should either say that the complexity is $\mathcal{O}(n^3)$ in practice, or choose beforehand a maximum number of Newton iteration that will be allowed in the algorithm.

6.3.6. An algorithm based on QR decomposition

The following algorithm synthetizes the procedure for the computation of an ϵ-GCD outlined so far in this section.

Algorithm PivQr

Input: polynomials $u(x)$ and $v(x)$, tolerance ϵ.
Output: an ϵ-GCD of $u(x)$ and $v(x)$; a backward error (residual of the GCD system); possibly perturbed polynomials $\hat{u}(x)$ and $\hat{v}(x)$ and cofactors $p(x)$ and $q(x)$.

1. Normalize $u(x)$ and $v(x)$.
2. Compute $M =$ either $S(u, v)$ or $B(u, v)$;
3. Perform QR decomposition with pivoting: $S = QR\Pi$;
4. Find k=upper bound on the ϵ-GCD degree using (6.3.7) or (6.3.10);
5. If $k = 0$, output "ϵ-coprime polynomials" and exit;
6. Solve the linear system given by (6.3.13), with output $p(x)$ and $q(x)$ (tentative cofactors);
7. Compute $g(x) = u(x)/p(x)$;
8. Perform iterative refinement on $g(x)$;

9. If $g(x)$ is an ϵ-divisor, output $g(x)$;
   else set $k = k - 1$ and go to step 5.

6.4. Conditioning

We collect here some remarks on the conditioning of the approximate GCD problem. Poor results due to ill-conditioning, however, are often greatly improved by iterative refinement.

6.4.1. Degree

In general, the degree of the ϵ-GCD of polynomials $p(x)$ and $q(x)$ is a nonconstant function of ϵ which takes integer values. Therefore, if seen as a real function, it is bound to be discontinuous. For example, take

$$p(x) = x - \alpha,$$
$$q(x) = x - \beta,$$
$$\epsilon = |\alpha - \beta|.$$

Then arbitrarily small perturbations of ϵ, α or β may change the degree of the ϵ-GCD from 0 to 1 or vice versa.

In practice, it seldom happens to find such an ill-conditioned situation, so in the following discussion it is reasonable to assume that there exist neighbourhoods of $p(x)$, $q(x)$ and ϵ where the degree of the ϵ-GCD is constant. Also observe that – as remarked in [74] – from a computational point of view, points of discontinuity are not sharply defined and should rather be understood as "critical ranges".

6.4.2. Coefficients

It should also be pointed out that, once the ϵ-GCD degree has been fixed, there is not a unique ϵ-GCD, but rather a set of ϵ-GCDs. However, it seems reasonable to ask how an ϵ-GCD changes upon perturbation of $p(x)$, $q(x)$ and ϵ.

Let $g(x)$ be an ϵ-GCD of $p(x)$, $q(x)$, i.e.,

$$\begin{cases} p(x) = g(x)u(x) + \epsilon_p(x) \\ q(x) = g(x)v(x) + \epsilon_q(x) \end{cases}$$

where $\|\epsilon_p(x)\| \leq \epsilon$, $\|\epsilon_q(x)\| \leq \epsilon$. The notation $\| \cdot \|$ is understood as the 2-norm, unless otherwise stated.

Assume that the ϵ-GCD is computed as in Algorithm PivQr, that is, by finding the cofactors at first, and subsequently performing a polynomial division. In both cases, linear least squares systems are being solved.

Perturbation theory for least squares problems provides the following well known result (see *e.g.* [70], Theorem 20.1):

Theorem 6.4.1. (Wedin) *Consider the linear least squares problem* $Ax = b$ *and assume that the perturbation on A and* $\mathbf{b}$ *is bounded in 2-norm by* δ. *Then the corresponding normalized perturbation of the solution satisfies*

$$\frac{\|\Delta \mathbf{x}\|_2}{\|\mathbf{x}\|_2} \leq \frac{\kappa \delta}{1 - \kappa \delta} \left(2 + (1 + \kappa) \frac{\|\mathbf{r}\|_2}{\|A\|_2 \|\mathbf{x}\|_2} \right),$$

where κ *is the 2-norm condition number of A and* $\mathbf{r}$ *is the residual (i.e.,* $\mathbf{r} = A\mathbf{x} - \mathbf{b}$).

Roughly speaking, this result says that the conditioning of the system is measured by κ if the residual is small, by κ^2 otherwise.

Now, denote with $\mathcal{C}$ the matrix of the cofactors system, whereas the unknown vector is $\mathbf{w} = [\mathbf{u}^T \quad \mathbf{v}^T]^T$. The residual for this least squares system is bounded in 2-norm by $\epsilon(\|\mathbf{u}\|_2 + \|\mathbf{v}\|_2)$. Therefore Wedin's theorem yields:

$$\frac{\|\Delta \mathbf{w}\|_2}{\|\mathbf{w}\|_2} \leq \frac{\delta \kappa_\mathcal{C}}{1 - \delta \kappa_\mathcal{C}} \left(2 + \left(1 + \kappa_\mathcal{C} \frac{\epsilon(\|\mathbf{u}\|_2 + \|\mathbf{v}\|_2)}{\|\mathcal{C}\|_2 \|\mathbf{w}\|_2} \right) \right) \quad (6.4.1)$$

$$\leq \frac{\delta \kappa_\mathcal{C}}{1 - \delta \kappa_\mathcal{C}} \left(2 + (1 + \kappa_\mathcal{C}) \epsilon \frac{\sqrt{2}}{\|\mathcal{C}\|_2} \right). \quad (6.4.2)$$

This result suggests that the condition number of $\mathcal{C}$ plays a key role in determining the conditioning of the problem. Moreover, we deduce that approximate GCD problems with a large tolerance tend to display worse condition properties than problems where a nearly exact GCD can be found.

The condition number of $\mathcal{C}$ can be seen as the inverse of the normalized distance between $\mathcal{C}$ and the nearest singular matrix. It follows that, up to multiplication for a constant depending on the size of the matrix, this condition number is greater than $1/\epsilon_0$, where

$$\epsilon_0 = \sup\{\eta : \deg \eta - \text{GCD}(p(x), q(x)) = k\}.$$

A similar remark can be applied to $\mathcal{C}_{u+v}$. Therefore, ill-conditioned GCD problems can roughly be seen as the cases in which the degree of the approximate GCD is particularly sensitive to the choice of the tolerance.

It should also be pointed out that, since the matrices that come up when studying GCD problems are usually structured (*e.g.*, block-Toeplitz), it would be more appropriate to replace the condition numbers mentioned in the above discussion with structured condition numbers. Structured condition numbers, however, are often difficult to compute and do not always give remarkable advantages (see [22] and [113]).

See also [144] for comments on GCD conditioning.

6.5. Tridiagonalization

We propose in this section an algorithm for ϵ-GCD computations which is based on the tridiagonalization of the Bézout matrix. This factorization is suggested by the symmetric structure of the matrix and, while not in general rank-revealing, it proves useful both in the determination of an upper bound on the ϵ-GCD degree and in the computation of coefficients. On a similar line of thought, see [103] for a brief discussion of the tridiagonalization of the Hankel matrix associated with a pair of polynomials.

6.5.1. The exact case

Let $B = B(u, v)$ be the Bezoutian matrix associated with two real univariate polynomials $u(x)$ and $v(x)$ of degrees respectively n and m. B is a real symmetric matrix and can therefore be reduced in tridiagonal form via Householder transformations. So we have:

$$T = HBH^T, \tag{6.5.1}$$

where H is a unitary matrix given by a product of Householder elementary matrices and T is tridiagonal symmetric.

Now, let $g(x)$ be the (exact) GCD of $u(x)$ and $v(x)$, and let $k = \deg g(x)$. If T is singular, it may have zero rows and columns, and/or a block diagonal structure with some singular blocks.

Remark 6.5.1. The first l rows and columns of T are zero $\Leftrightarrow$ the first l rows and columns of B are zero $\Leftrightarrow$ 0 is a common root for $u(x)$ and $v(x)$ with multiplicity l.

Proof. It is obvious that the first l tridiagonalization steps send zero columns and rows to zero columns and rows. On the other hand, if the first l rows and columns of T are zero, the same must be true for B, since H is unitary and therefore nonsingular. This proves the first equivalence. The second one is easily verified, *e.g.* from the recursive formula for the Bezoutian (see Proposition 2.6.4). $\qquad\square$

So we can rule out the case when the first rows and columns of T are zero by requiring that the polynomials we are working with do not have zero common roots.

In an approximate setting, this can be achieved either using the Bezoutian matrix of the reversed polynomials, or detecting possible zero roots by checking if 0 belongs to the root neighborhoods of both polynomials.

In the following, the first row and column of T will always be assumed to be nonzero. We will say that the block diagonal structure of T is *nontrivial* if the upper-left block does not consist of the whole matrix (see *e.g.* the matrix T in Example 6.5.5), and *strongly non-trivial* if there are nonzero entries outside the upper-left block (see *e.g.* the matrix T in Example 6.5.6).

Remark 6.5.2. If T is singular, then the upper-left block of its (possibly trivial) block-diagonal structure is singular.

Proof. If T is singular, then $u(x)$ and $v(x)$ are not coprime. Let α be a common root for $u(x)$ and $v(x)$. The vector $\mathbf{w} = [1 \quad \alpha \quad \alpha^2 \ldots \alpha^{n-1}]^T$ belongs to the null space of $B(u, v)$ and therefore $\mathbf{z} = H\mathbf{w}$ belongs to the null space of T. Notice that the first entry of $\mathbf{z}$ is 1, because the first row of H is $[1 \quad 0 \ldots 0]$. Let T_s be the upper-left block of T and let s be its size; then $\mathbf{z}(1 : s)$ is a nonzero vector such that $T_s\mathbf{z}(1 : s) = 0$. □

Theorem 6.5.3. *Let $T = HB(u, v)H^T$ be the Householder tridiagonalization of $B(u, v)$, where $u(x)$ and $v(x)$ are real polynomials with no zero common roots and having a nontrivial GCD of degree k. Then for almost any choice of $u(x)$ and $v(x)$, the tridiagonal matrix T can be split as the direct sum of a singular irreducible $(n - k) \times (n - k)$ tridiagonal matrix and a null $k \times k$ matrix.*

Proof. Assume $k \geq 2$. We want to show that for almost any choice of $u(x)$ and $v(x)$ the last $k - 1$ rows and columns of T are zero. If this is not the case, then T must have a strongly non-trivial block diagonal structure. Now, observe that the Householder tridiagonalization can be seen as a Lanczos tridiagonalization process, with starting vector given by the first column of H, i.e., $[1 \quad 0 \ldots 0]^T$. Lanczos tridiagonalization is known to produce a strongly non-trivial block diagonal structure when the starting vector lies in a proper invariant subspace of $B(u, v)$. The set of pairs (u, v) such that $[1 \quad 0 \ldots 0]^T$ lies in a proper invariant subspace of the Bezoutian has zero measure; hence the thesis. The case $k = 1$ is easily proved using a similar argument. □

Remark 6.5.4. Since $B(u, v)$ is a real symmetric matrix and therefore admits an orthogonal basis of eigenvectors, it follows that $[1 \quad 0 \dots 0]$ lies in a proper invariant subspace of $B(u, v)$ iff at least one of the eigenvectors associated with nonzero eigenvalues has 0 as first entry.

Proposition 6.5.3 states that the generic form of T has an unreduced tridiagonal upper-left block followed by zero rows and columns. However, counterexamples to this behaviour do exist, as shown in the examples below.

Example 6.5.5. (Generic form) Let

$$u(x) = x^6 + 2x^5 - 6x^4 + 31x^3 - 7x^2 + x + 6$$
$$= (x^3 - 2x^2 + 5x + 3)(x^3 + 4x^2 - 3x + 2),$$
$$v(x) = x^6 - 4x^5 + 18x^4 - 20x^3 + 29x^2 + 52x + 15$$
$$= (x^3 - 2x^2 + 5x + 3)(x^3 - 2x^2 + 9x + 5).$$

The associated Bezoutian and its tridiagonal form are:

$$B(u, v) = \begin{pmatrix} -297 & -279 & 585 & -198 & 54 & 9 \\ -279 & 192 & 1434 & -276 & 117 & 51 \\ 585 & 1434 & 483 & 69 & 81 & 36 \\ -198 & -276 & 69 & -357 & 120 & -51 \\ 54 & 117 & 81 & 120 & -39 & 24 \\ 9 & 51 & 36 & -51 & 24 & -6 \end{pmatrix},$$

$$T = 10^3 \begin{pmatrix} -0.2970 & 0.6799 & 0 & 0 & 0 & 0 \\ 0.6799 & -0.7552 & 1.0477 & 0 & 0 & 0 \\ 0 & 1.0477 & 1.3691 & 0.0177 & 0 & 0 \\ 0 & 0 & 0.0177 & -0.3409 & 0 & 0 \\ 0 & 0 & 0 & 0 & 0 & 0 \\ 0 & 0 & 0 & 0 & 0 & 0 \end{pmatrix}.$$

(nonzero entries of T are rounded to four decimal digits).

Example 6.5.6. (Counterexample to generic form) Let

$$u(x) = -2x^3 + 3x^2 - 2x + 1 = (x - 1)(-2x^2 + x - 1),$$
$$v(x) = -3x^3 + 2x^2 + 1 = (x - 1)(-3x^2 - x - 1).$$

The associated Bezoutian and its tridiagonal form are:

$$B(u, v) = \begin{pmatrix} -2 & 1 & 1 \\ 1 & 5 & -6 \\ 1 & -6 & 5 \end{pmatrix}, \qquad T = \begin{pmatrix} -2 & \sqrt{2} & 0 \\ \sqrt{2} & -1 & 0 \\ 0 & 0 & 11 \end{pmatrix}.$$

Observe that T has a strongly non-trivial block diagonal structure and T_2 is a singular block. Also notice that $[0 \quad 1 \quad -1]^T$ is an eigenvector of $B(u, v)$ with eigenvalue 11.

Example 6.5.7. (another counterexample to generic form) Let

$$u(x) = x^4 - 1,$$
$$v(x) = u'(x) = 4x^3.$$

The associated Bezoutian and its tridiagonal form are:

$$B(u, v) = \begin{pmatrix} 0 & 0 & 4 & 0 \\ 0 & 4 & 0 & 0 \\ 4 & 0 & 0 & 0 \\ 0 & 0 & 0 & 4 \end{pmatrix}, \qquad T = \begin{pmatrix} 0 & 4 & 0 & 0 \\ 4 & 0 & 0 & 0 \\ 0 & 0 & 4 & 0 \\ 0 & 0 & 0 & 4 \end{pmatrix}.$$

In the proof of Theorem 6.5.3 it is pointed out that $B(u, v)$ tridiagonalizes to a nongeneric form iff $[1 \quad 0 \ldots 0]^T$ lies in a proper invariant subspace of $B(u, v)$. In some cases, this difficulty can easily be solved:

Remark 6.5.8. Suppose that $B(u, v)$ tridiagonalizes to a non-generic form and assume that there exists a vector $\mathbf{y}$ that does not lie in a proper invariant subspace of $B(u, v)$. (Under these hypotheses, a random choice yields such a vector with probability 1.) Let P be the Householder matrix associated with $\mathbf{y}$. Then the Householder tridiagonalization of PBP^T gives a matrix

$$T = \tilde{H}(PBP^T)\tilde{H}^T$$
$$= (\tilde{H}P)B(\tilde{H}P)^T$$

of generic form.

However, there are Bezoutian matrices – such as the one in Example 6.5.7 – for which any vector lies in a proper invariant subspace.

Lemma 6.5.9. *Let $S \in \mathbb{R}^{n \times n}$ be a symmetric matrix. Then S has at least a nonzero eigenvalue of multiplicity greater than 1 if and only if, for every vector $\mathbf{y} \in \mathbb{R}^n$, $\mathbf{y}$ lies in a proper invariant subspace of S.*

Proof. Without loss of generality, assume that S is nonsingular. Observe that the existence of a vector $\mathbf{z} \in \mathbb{R}^n$ that does not belong to any proper invariant subspace of S is equivalent to the existence of a vector $\mathbf{w} \in \mathbb{R}^n$

such that the set $\mathbf{w}, S\mathbf{w}, \ldots, S^{n-1}\mathbf{w}$ is linearly independent; that is, such that the matrix

$$\Lambda = [\mathbf{w} \quad S\mathbf{w} \quad \ldots \quad S^{n-1}\mathbf{w}]$$

is nonsingular.

Let $(\lambda_i, \mathbf{v}_i)_{i=1,\ldots n}$ be the eigenvalues and eigenvectors of S. If a vector $\mathbf{w}$ as above exists, then it can be written as

$$\mathbf{w} = \sum_{i=1}^{n} \alpha_i \mathbf{v}_i, \qquad (6.5.2)$$

where $\alpha_i \neq 0$ for $i = 1, \ldots n$. Indeed, if there were an index i_0 for which $\alpha_{i_0} = 0$, then $\mathbf{w}$ would belong to the invariant subspace $< \mathbf{v}_{i_0} >^{\perp}$.

It follows from (6.5.2) that

$$A^j \mathbf{w} = \sum_{i=0}^{n} \alpha_i \lambda_i^j \mathbf{v}_i$$

and therefore Λ can be written as

$$\Lambda = \begin{bmatrix} \alpha_1 & \alpha_1\lambda_1 & \ldots & \alpha_1\lambda_1^{n-1} \\ \alpha_2 & \alpha_2\lambda_2 & \ldots & \alpha_2\lambda_2^{n-1} \\ \vdots & \vdots & & \vdots \\ \alpha_n & \alpha_n\lambda_n & \ldots & \alpha_n\lambda_n^{n-1} \end{bmatrix}.$$

It follows that Λ is a Vandermonde matrix and its determinant is zero if and only if there exist indices $i \neq j$ such that $\lambda_i = \lambda_j$. $\qquad \square$

So, if our Bezoutian matrix has an eigenvalue of multiplicity 2 or greater, it cannot be tridiagonalized to a generic form. Unless otherwise stated, we will assume in the following that $B(u, v)$ tridiagonalizes to a generic form. It will be pointed out later how Bezoutians with nongeneric tridiagonal form can be dealt with.

Remark 6.5.10. Suppose that T is singular and let l be the number of zero rows and columns in T, which, by our hypotheses, are to be found at the "end" of T, and let $\hat{T}$ be the tridiagonal matrix given by T deprived of its zero rows and columns. It follows from Remark 6.5.2 that $\hat{T}$ is singular. Then rank $B(u, v) =$ rank $\hat{T} = n - l - 1$.

Besides being useful for rank determination, tridiagonalizing $B(u, v)$ enables us to calculate the coefficients of GCD(u, v).

Indeed, suppose that $l > 0$. From the factorizations (6.5.1) and (2.6.4) we have:

$$T = HG\hat{B}G^T H^T. \qquad (6.5.3)$$

The last l rows and columns of this matrix are zero. Since G, $\hat{B}$ and H have maximum rank, it follows that the last l rows of HG must be zero. Let $h = [h_1 \ldots h_n]$ be a row vector such that $hG = [0 \ldots 0]$. Such a condition can be expressed through the following Hankel linear system:

$$\begin{cases} h_1 g_0 + h_2 g_1 + \ldots + h_{k+1} g_k = 0 \\ h_2 g_0 + h_3 g_2 + \ldots + h_{k+2} g_k = 0 \\ \cdots\cdots\cdots\cdots\cdots\cdots\cdots\cdots \\ h_{n-k} g_0 + h_{n-k+1} g_1 + \ldots + h_n g_k = 0 \end{cases}$$

Since we would like the GCD to be monic, we can assume $g_k = 1$; the above linear system becomes

$$A\hat{g} = b, \tag{6.5.4}$$

where $\hat{g} = [g_0 \ldots g_{k-1}]^T$ is a vector containing the coefficients of $g(x)$ except for the leading one, $p = -[h_{k+1} \ldots h_n]^T$ and

$$A = \begin{pmatrix} h_1 & h_2 & \ldots & h_{k+1} \\ h_2 & h_3 & \ldots & h_{k+2} \\ \vdots & \vdots & & \vdots \\ h_{n-k} & h_{n-k+1} & \ldots & h_n \end{pmatrix}.$$

Each of the last l rows of H, which we will call h_i, with $i = n - l + 1, \ldots, n$, gives a linear system $A_i \hat{g} = b_i$ built like (6.5.4). Besides, an additional vector in the null space of T is easily computed from $\hat{T}$, and it yields a system of the type (6.5.4) as well. So we obtain a system

$$K\hat{g} = p, \tag{6.5.5}$$

where

$$K = \begin{bmatrix} A_{n-l+1} \\ A_{n-l+2} \\ \vdots \\ A_n \end{bmatrix} \quad \text{and} \quad p = \begin{bmatrix} b_{n-l+1} \\ b_{n-l+2} \\ \vdots \\ b_n \end{bmatrix},$$

which is basically derived from a well-conditioned (in fact, orthogonal) set of generators for the null space of B.

Solving (6.5.5) yields the coefficients of $g(x)$.

Suppose now that $l = 0$ but T is still singular, which implies $k = 1$. Then $u(x)$ and $v(x)$ have only one common root α. It follows from Remark 2.6.10 that $\mathrm{Ker} B(u, v)$ is spanned by $[1, \alpha, \alpha^2, \ldots, \alpha_i^{n-1}]^T$. So, if a vector $v \in \mathrm{Ker} B(u, v)$ is known, then α can be calculated as the quotient of two consecutive entries of v. But such a vector can be retrieved from a vector $w \in \mathrm{Ker} T$, which is easily calculated, as $v = H^H w$.

Finally, let us examine the case when T is nongeneric. The above method cannot be directly applied because there is no guarantee that the last (dim ker $T - 1$) rows of T should be zero; so we might need to find more independent vectors in the null space of T. This is done by checking whether each block of T is singular and computing a vector in its null space, if possible.

6.5.2. The approximate case: rank determination

The above discussion holds in the exact case, but the stability of Householder tridiagonalization suggests that it can be extended to an approximate setting. Roughly speaking, one may proceed as follows:

1. Given polynomials $u(x)$ and $v(x)$, compute $B = B(u, v)$;
2. Perform Householder tridiagonalization on B, yielding $T = HBH^H$;
3. Determine (an upper bound on) the degree of an approximate GCD;
4. Compute the coefficients of an approximate GCD of $u(x)$ and $v(x)$ by solving the linear system (6.5.5) obtained from the last "small" rows of H.

Step 3 needs to be described more carefully. If a tolerance ϵ is given, how can we detect the degree of an ϵ-GCD from T? An upper bound can be found as follows.

Let $u(x)$, $v(x)$, $\hat{u}(x)$ and $\hat{v}(x)$ be as in Lemma 4.1.3. Suppose that $\hat{u}(x)$ and $\hat{v}(x)$ have an ϵ-GCD of degree d, which we may view as the exact GCD of $u(x)$ and $v(x)$. Then the matrix $B(u, v)$ has rank $n - d$. If $\sigma_1, \ldots, \sigma_n$ are the singular values of $B = B(\hat{u}, \hat{v})$, Theorem 4.1.2 says that σ_{n-d+1} is the 2-norm distance between B and the closest matrix of rank $n - d$. It follows that

$$\sigma_{n-d+1} \leq \| B(u, v) - B(\hat{u}, \hat{v}) \|_2$$

and applying Lemma 4.1.3 we have

$$\sigma_{n-d+1} \leq 4n\epsilon. \tag{6.5.6}$$

It follows from inequality (6.5.6) that if the singular values of B are known, then

$$k_0 \equiv \max\{k \in \mathbb{N} : \sigma_{n-k+1} \leq 4n\epsilon\}$$

is an upper bound for the degree of an ϵ-GCD of $\hat{u}(x)$ and $\hat{v}(x)$.

In the real case, the singular values of B coincide with the absolute values of its eigenvalues, since B is real symmetric. But we have already computed a tridiagonal form of B, whose eigenvalues are easily found.

6.5.3. The algorithm

Algorithm TdBez for approximate GCD

Input: Real polynomials $u(x)$ and $v(x)$ of degrees n and m, with $n \geq m$; tolerance ϵ.

Output: An ϵ-GCD for $u(x)$ and $v(x)$; a backward error (residual of the GCD system); possibly perturbed polynomials $\hat{u}(x)$ and $\hat{v}(x)$ and cofactors $p(x)$ and $q(x)$.

1. Normalize $u(x)$ and $v(x)$ so that they have unitary 2-norm.
2. Calculate the permuted Bezoutian $B = JB(u, v)J$.
3. Tridiagonalize $B(u, v)$ using Householder transformations, thus obtaining a tridiagonal matrix $T = HBH^H$ of generic form, if possible.
4. Calculate an upper bound k_0 on the degree of an ϵ-GCD.
5. Set $k = k_0$.
6. If $k = 0$, output 'Coprime polynomials';
   else if $k = 1$, find a vector $\mathbf{w} = [w_1 \ldots w_n] \in \ker T$ and set $g(x) = x - \frac{w2}{w1}$;
   else if $k = m$, set $g(x) = v(x)$;
   else solve the system (6.5.5) obtained from the last $k - 1$ rows of H (if T is in generic form, otherwise compute vectors in the null space of each block) and a vector in $\mathrm{Ker}\hat{T}$ and set $g(x) = x^k + \sum_{i=0}^{k-1} g_i x^i$.
7. Perform iterative refinement on $g(x)$.
8. If $g(x)$ is an ϵ-divisor, output $g(x)$;
   else set $k = k - 1$ and repeat steps 6-8.

6.6. More factorization-based algorithms

We briefly describe here three more algorithms for the computation of an ϵ-GCD of given polinomials $u(x)$ and $v(x)$. The first two methods are based on the SVD and on the QR factorization of the Bézout matrix, whereas the third one relies on the QR factorization of another resultant matrix, namely the matrix $v(F_u)$ defined in Section 2.7.

For ease of exposition, ϵ is assumed to be the threshold used to determine the "correct" approximate rank of the resultant matrix. The modifications required so that the algorithms actually compute an ϵ-GCD are straightforward and follow the pattern of algorithms PivQr and TdBez. Moreover, iterative refinement may be added.

6.6.1. SVD of the Bézout matrix

This method employs the SVD of the Bezoutian both for approximate rank determination and for the computation of coefficients of an approximate GCD. A QR factorization is also required.

The approximate GCD degree is determined through Lemma 4.1.4. The computation of the coefficients is based on the following discussion (which applies to the exact case and assumes that $u(x)$ and $v(x)$ have an exact GCD $g(x)$).

Recall from Proposition 2.7.3 that the Bezoutian associated with $u(x)$ and $v(x)$ factorizes as

$$B(u, v) = B(u, 1)H(u, v)B(u, 1).$$

Therefore, if $\mathbf{r} \in \ker B(u, v)$, it follows that $\mathbf{s} = B(u, 1)\mathbf{r}$ belongs to the null space of $H(u, v)$.

The null space of $H(u, v)$ is characterized in Proposition 2.7.5, which, in particular, says that the cofactor $p(x) = u(x)/g(x)$ can be written as a linear combination of generators of $\ker H(u, v)$.

Recall from Section 4.1 that the SVD provides an orthogonal basis for the null space of the factorized matrix. So we can extract from the SVD of $B(u, v)$ a matrix C whose rows are, when transposed, an orthogonal basis for $B(u, v)$. Now, perform the QR factorization of C; the last row $\mathbf{r}_*^T$ is then the "shortest" (i.e., with the maximum number of prefixed zeros) vector that can belongs to $\ker B(u, v)$. Moreover, $\mathbf{s}_* = B(u, 1)\mathbf{r}_*$ is the "shortest" vector in $\ker H(u, v)$. Let $s(x)$ be the associated polynomial; then it follows that $g(x) = u(x)/s(x)$ is a GCD of $u(x)$ and $v(x)$.

Algorithm KerSVD

Input: Polynomials $u(x)$ and $v(x)$ of degrees n and m, with $n \geq m$; a tolerance ϵ.

Output: An approximate GCD for $u(x)$ and $v(x)$.

- Compute $B(u, v)$.
- Perform the SVD of $B(u, v)$ and find U, V, Σ such that $B(u, v) = U\Sigma V^H$.
- Let $k =$ number of singular values greater than ϵ.
- Let $C = V(k + 1 : n)^T$; the rows of C span the null space of $B(u, v)$.
- Perform the QR decomposition of C; let $\mathbf{r}^T$ be the last row of the R factor.
- Let $s(x)$ be the polynomial associated with $B(u, 1)\mathbf{r}$ (i.e., having its entries as coefficients).
- Let $g(x) = u(x)/s(x)$.
- Output $g(x)$.

6.6.2. QR decomposition of the Bézout matrix

The method for the computation of the approximate GCD proposed in [41] relies on the fact that the polynomial GCD can be retrieved from the last nonzero row in the triangular factor given by the QR factorization of the Sylvester matrix.

We have pointed out (see Theorem 2.6.11) that a similar property also holds for the Bézout matrix. Therefore it is possible to design an approximate GCD algorithm based on the QR factorizarion of the Bezoutian. Moreover, experimental evidence seems to show that the QR factorization of the Bezoutian is less prone than the factorization of the Sylvester matrix to the instability issues caused by the presence of large common roots; this property makes the root separation task much easier. The approximate GCD can be seen as the product of two polynomials $g_1(x)$ and $g_2(x)$, where $g_1(x)$ is determined from the QR factorization of the permuted Bezoutian, whereas $g_2(x)$ is computed from the QR factorization of the nonpermuted Bezoutian (that is, as a linear combination of the reversed input polynomials). Roughly speaking, an approximate GCD algorithm might go as follows:

Algorithm BezQR

Input: Polynomials $u(x)$ and $v(x)$ of degrees n and m, with $n \geq m$; a tolerance ϵ.
Output: An approximate GCD for $u(x)$ and $v(x)$.

- Compute $\hat{B} = J B(u, v) J$ (where J is a permutation matrix as in Theorem 2.6.11).
- Factorize $\hat{B} = QR$.
- Let $\mathbf{r}^T$ be the last row of R with norm larger than ϵ.
- Extract from $\mathbf{r}^T$ the coefficients of an approximate divisor $g_1(x)$ of $u(x)$ and $v(x)$.
- Compute $u_1(x) = u(x)/g_1(x)$ and $v_1(x) = v(x)/g_1(x)$.
- Compute $B_1 = B(u_1, v_1)$.
- Factorize $B_1 = Q_1 R_1$.
- Let $\mathbf{r}_1^T$ be the last row of R with norm larger than ϵ.
- Extract from $\mathbf{r}_1^T$ the coefficients (which are in reversed order this time) of an approximate divisor $g_2(x)$ of $u(x)$ and $v(x)$.
- Output $g(x) = g_1(x) \cdot g_2(x)$.

6.6.3. QR decomposition of the companion matrix resultant

Recall that the the matrix $v(F_u)$ defined in Section 2.7 has properties very similar to the Sylvester and Bézout matrices and qualifies as a resultant matrix. We exploit these properties in an algorithm for the computation of an approximate GCD. The same conventions for the description of the algorithm that have been used in the previous section apply here.

Actually, we are not using here the matrix $v(F_u)$ but rather its transpose $W = v(F_u^T)$. Let $W = QR$ be the QR factorization of W. Then it follows from Theorem 2.7.1 and from the symmetry of the Bezoutian that

$$B(u, v) = B(u, v)^T = W B(u, 1) = QRB(u, 1)$$

from which we obtain

$$Q^T B(u, v) = RB(u, 1).$$

Recall that $B(u, 1)$ is an upper left triangular matrix. Now, let $\mathbf{r}^T$ be the last nonzero row of R. Then $\mathbf{g} = \mathbf{r}^T B(u, 1)$ is the last nonzero row of $RB(u, 1)$. The polynomial $g(x)$ associated with the vector $\mathbf{g}$ has the same degree as the GCD of $u(x)$ and $v(x)$. Moreover, the vector of its coefficients is obtained as a linear combination of the rows of $B(u, v)$. Therefore $g(x)$ is a GCD.

The adaptation to the approximate case goes as usual.

Algorithm QRres

Input: Polynomials $u(x)$ and $v(x)$ of degrees n and m, with $n \geq m$; a tolerance ϵ.
Output: An approximate GCD for $u(x)$ and $v(x)$.

- Compute $W = v(F_u^T)$.
- Factorize $W = QR$.
- Let $\mathbf{r}^T$ be the last row of R with norm larger than ϵ.
- Compute $\mathbf{g} = \mathbf{r}^T B(u, 1)$.
- Output the polynomial $g(x)$ having the entries of $\mathbf{g}$ as coefficients.

The interesting features of this algorithm are the use of the structured resultant matrix $v(F_u)$ and the use of the QR factorization, which might potentially be implemented using a fast algorithm.

Stability issues need to be considered, since the entries of $v(F_u)$ might grow quite large. However, it should be pointed out that $v(F_u)$ is actually a Bezoutian expressed in the Horner (or control) basis (see [29] and [69]) and may therefore represent a useful tool for the study of polynomial GCD in bases different from the usual monomial one.

See also Section 9.5 for more results on GCD computation via the companion matrix resultant.

Chapter 7
A fast GCD algorithm

The algorithms for approximate GCD computation outlined in the previous chapters all have a computational cost which is cubic in the degrees of the polynomials. The only exception is the Euclidean algorithm, which however, in addition to troublesome stability issues, might fail to return an approximate divisor of maximum degree.

But since GCD computations deal with structured matrices, it is conceivable that algorithms designed to exploit the structure (namely, the displacement structure) might require less computational effort. Fast (*i.e.*, $\mathcal{O}(n^2)$) algorithms for the QR or LU factorization of displacement structured matrices are indeed available (see *e.g.* [79, 76, 106]) and can be applied to the solution of the approximate GCD problem. Efforts in this direction are found for instance in [146], where fast QR factorization is applied to the Sylvester matrix.

Fast methods, though, often display poor stability properties, and so do the approximate GCD methods that are derived from them. Here we propose a fast algorithm for the computation of an approximate GCD which is also quite stable. The method relies on a stabilized version of fast LU factorization of the Sylvester or Bézout matrix for the determination of degree and coefficients of an ϵ-GCD and for iterative refinement. See also [13].

7.1. Stability issues

Gaussian elimination with partial pivoting (GEPP) is usually regarded as a fairly reliable method for solving linear systems. How does the GKO algorithm (see Section 2.4) perform, from the point of view of numerical stability, when compared to ordinary GEPP?

Standard error analysis of GEPP applied to an arbitrary $n \times n$ matrix M shows that the backward error essentially depends on $\|\hat{U}\|_\infty$, where $\hat{U}$ is the computed upper triangular factor in the LU factorization of M. In other words, if the entries of $\hat{U}$ do not grow too large, than the method is stable. The theoretical bound

$$\|U\|_{\max}/\|C\|_{\max} \leq 2^{n-1}$$

is known to be sharp, but very rarely attained in practice.

Since the GKO algorithm yields exactly the same LU factorization of a Cauchy-like matrix as GEPP, element growth in the upper triangular factor may represent a source of instability.

Sweet and Brent ([125]) have done an error analysis of the GKO algorithm applied to a Cauchy-like matrix C. They point out that the error propagation depends not only on the magnitude of the triangular factors in the LU factorization of C (as it is expected for ordinary Gaussian elimination), but also on the magnitude of the generators. In some cases, the generators can suffer from large internal growth, even if the triangular factors do not grow too large, and therefore cause a corresponding growth in the backward and forward error.

However, it is possible to modify the GKO algorithm so as to prevent generator growth, as suggested for example in [87] and [123]. We describe here a variation of GECP (Gaussian elimination with complete pivoting) proposed by Ming Gu ([87]).

This method basically involves orthogonalization of the first generator at each elimination step. This guarantees that the first generator is well conditioned and allows a good choice of a pivot. Each step of the GKO algorithm is modified as follows:

- Orthogonalize the first generator G, *i.e.*
    - QR-factorize G, obtaining $G = \mathcal{G}R$, where $\mathcal{G}$ is an $n \times \alpha$ column orthogonal matrix and R is upper triangular;
    - define new generators $\tilde{G} = \mathcal{G}$ and $\tilde{B} = RB$.
- Determine the index h of the column of maximum 2-norm in B.
- Let P_2 be the (right) permutation matrix that interchanges the first and h-th columns. Interchange the first and h-th diagonal entries of F_2; interchange the first and h-th columns of B.
- Compute the first column of $C \cdot P_2$ from the generators and determine the position (say, $(k, 1)$) of the entry of maximum magnitude.
- Let P_1 be the permutation matrix that interchanges the first and k-th rows. Interchange the first and k-th diagonal entries of F_1; interchange the first and k-th rows of G.
- Recover from the generators the first row of $P_1 \cdot C$. Now one has the first column of L and the first row of U in the LU factorization of $P_1 \cdot C \cdot P_2$.
- Compute generators of the Schur complement C_2 of $P_1 \cdot C \cdot P_2$.

In other words, this method performs partial pivoting on the column of C corresponding to the column of B with maximum norm. This technique is not equivalent to complete pivoting, but nevertheless allows a good choice of pivots and effectively reduces element growth in the generators, as well as in the triangular factors. Indeed, define

$$\xi_{max} = \max_{i,j} |D_1(i, i) - D_2(j, j)|, \quad \xi_{min} = \min_{i,j} |D_1(i, i) - D_2(j, j)|,$$

and $\rho = \dfrac{\xi_{max}}{\xi_{min}}.$

(Observe that for a Cauchy-like matrix derived from a Toeplitz-like one has $\rho = 2 \sin(\frac{\pi i}{2n})$.) Then we have (see again [87]):

Lemma 7.1.1. *Let C be a Cauchy-like matrix as in Section 2.2.2 and denote with j_{max} the index of the column with largest 2-norm in $G \cdot B$. Then*

$$\|C\|_{max} \lesssim \sqrt{n}\rho\|C_{:,j_{max}}\|_\infty$$

and

$$\|C\|_F \leq n\rho\|C_{:,j_{max}}\|_\infty.$$

If the first generator G is column orthogonal, then j_{max} is also the index of the column with maximum 2-norm in B, and it is therefore easily determined. Moreover, one has

$$\||G_{k:n,:}| \cdot |B_{:,k:n}|\|_2 \leq \||G_{k:n,:}|\|_2 \cdot \||B_{:,k:n}|\|_2$$
$$\leq \|G_{k:n,:}\|_F \cdot \|B_{:,k:n}\|_F \leq \sqrt{\alpha}\|B_{:,k:n}\|_F$$
$$\leq \alpha\|B_{:,k:n}\|_2 = \alpha\|G_{k:n,:} \cdot B_{:,k:n}\|_2,$$

thus avoiding internal generator growth.

Remark 7.1.2. Error analysis for Gu's algorithm shows that the backward error essentially depends on ρ and on the magnitude of the computed upper triangular factor (but not on the magnitude of the generators).

More precisely, let $C = \hat{L}\hat{U} + H$, where $\hat{L}$ and $\hat{U}$ are the factors of the LU factorization of C computed using the reorthogonalization technique, and H is the backward error matrix. Then the following bound is proved in [87]:

$$\|H\|_\infty \leq 8\sqrt{\alpha}\bar{\eta}(\alpha + 2) \cdot n^2 \cdot \rho \cdot \left(\|\hat{U}\|_\infty + n\mu\right) + \mathcal{O}(\epsilon^2), \qquad (7.1.1)$$

with $\bar{\eta}$ a small multiple of ϵ and

$$\mu = \max_{2 \leq k \leq n} \|\hat{C}^{(k)}\|_{max},$$

where $\hat{C}^{(k)}$ is the factorized matrix at the k-th elimination step:

$$\hat{C}^{(k)} = \begin{pmatrix} 1 & 0 \\ \ell^{(k)} & I \end{pmatrix} \cdot \begin{pmatrix} d_k & u^{(k)} \\ 0 & C^{(k+1)} \end{pmatrix}.$$

Notice, however, that in some cases (*e.g.* when factorizing Cauchy-like matrices than are obtained from Toeplitz-plus-Hankel matrices) this bound may grow quite fast with n.

The key idea in stabilized versions of GKO is to modify the generators so as to keep them well-conditioned. In Gu's work is it pointed out that, for practical purposes, it might be enough to orthogonalize only every K steps, where K is a user-defined parameter ($K = 10$ seems to be a reasonable choice), so that the method becomes less expensive. We might also employ Gram-Schmidt orthogonalization in place of QR. Stewart suggests instead to perform LU factorization on the first generator ([123]), with a different pivoting scheme. See also [4] for a Matlab implementation of GKO equipped with several pivoting strategies.

7.2. Fast and stable factorization of rectangular matrices

Numerical experiments show that the GKO algorithm applied to a Sylvester matrix is often unstable because of generator growth.

Example 7.2.1. Two polynomials $f(x)$ and $g(x)$, of degrees respectively 54 and 53, are chosen as in Section 8.3. The polynomials have an exact GCD of degree 50. We compute the (horizontal) Sylvester matrix S associated with $f(x)$ and $g(x)$, and the Cauchy-like matrix C obtained from S. Fast GEPP is then applied to C. It turns out that the norm of the second generator grows very large in the process, as shown in Figure 7.1.

The backward error for this fast LU factorization of C is about $4.5 \cdot 10^{-7}$, and the computed U factor is of little use from the point of view of GCD computations, because it shows no gap in row magnitudes (see Figure 7.3).

Experimental evidence shows that GKO modified with orthogonalization of the first generator and selection of the column of maximum 2-norm from the second generator is quite stable and gives results comparable to GECP, whereas Stewart's method does not seem to be as successful for Sylvester matrices.

Therefore it seems best to stabilize the GKO algorithm by orthogonalizing the first generator and selecting the column of maximum 2-norm from the second generator. For ease of reference, we will call this pivoting technique *almost complete pivoting*, in view of Lemma 7.1.1.

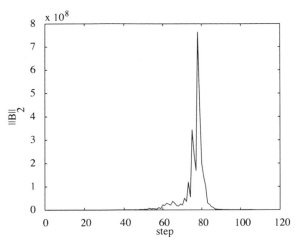

Figure 7.1. Norm of the second generator at each step of GKO.

Example 7.2.2. Fast LU factorization with almost complete pivoting applied to the same matrix of Example 7.2.1 gives satisfying results. The norm of both generators do not grow very large (Figure 7.2 shows the behaviour of the second generator) The backward error for the factorization of C is about $3.2 \cdot 10^{-13}$, and the U factor distinctly shows a gap in magnitude between the 57th and the 58th row (see Figure 7.4).

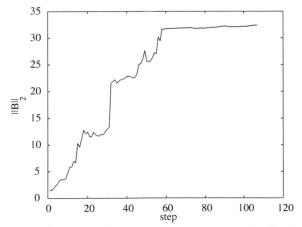

Figure 7.2. Norm of the second generator at each step of the fast LU factorization with almost complete pivoting.

We will also need to generalize fast LU to rectangular Toeplitz-like matrices, so that it can be applied to Sylvester subresultants. The trouble with a straightforward generalization of Proposition 2.4.2 to an $n \times m$ matrix is that the spectra of the matrices D_1 and D_2 associated with the

resulting Cauchy-like matrix might not be well separated. As a consequence, denominators in (2.2.7) might be very small, or even zero. This
is bound to happen, for example, if $n = 2m$.

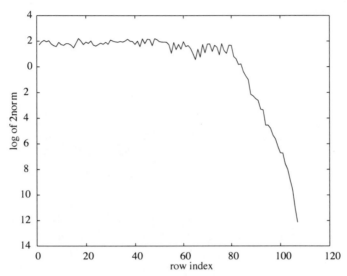

Figure 7.3. Norm of the rows of the computed U with GKO.

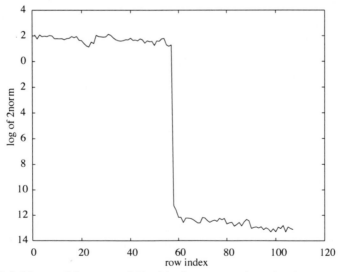

Figure 7.4. Norm of the rows of U with almost complete pivoting.

However, we can use an equivalent definition of Toeplitz matrix to obtain a Cauchy-like matrix such that the spectrum of the associated D_2 is
rotated of an arbitrary angle.

Proposition 2.4.2 exploits the fact that, if T is an $m \times m$ Toeplitz-like matrix, then the matrix

$$\nabla_{Z_1, Z_{-1}}(T) = Z_1 T - T Z_{-1}$$

has low rank α. An equivalent characterization of Toeplitz-like matrices can be obtained by requiring that the matrix

$$\nabla_{Z_1, Z_\theta}(T) = Z_1 T - T Z_\theta$$

should have low rank, for an arbitrary $\theta \in \mathbb{C}$ such that $|\theta| = 1$. Let $N = \mathrm{lcm}(n, m)$. The matrix Z_θ factorizes as

$$Z_\theta = D\mathcal{F}D_\theta\mathcal{F}^*D^{-1},$$

where

$$D_\theta = \theta \cdot D_1, \tag{7.2.1}$$

$$D = \mathrm{diag}(1, e^{\frac{\pi i}{Nm}}, e^{\frac{2\pi i}{Nm}}, \dots). \tag{7.2.2}$$

Therefore, the following generalized version of Proposition 2.4.2 holds:

Proposition 7.2.3. *Let $T \in \mathbb{C}^{n \times m}$ be a Toeplitz-like matrix, satisfying*

$$\nabla_{Z_1, Z_\theta}(T) = Z_1 T - T Z_\theta = G \cdot B,$$

where $G \in \mathbb{C}^{n \times \alpha}$, $B \in \mathbb{C}^{\alpha \times m}$ and Z_1, Z_θ are as in (2.2.6). Then

$$C = \mathcal{F}_n T D_\theta \mathcal{F}_m$$

is a Cauchy-like matrix, i.e.

$$\nabla_{D_1, D_\theta}(C) = D_1 C - C D_\theta = \hat{G}\hat{B}, \tag{7.2.3}$$

where $\mathcal{F}_n$ and $\mathcal{F}_m$ are the normalized Discrete Fourier Transform matrices of order n and m respectively, D and D_θ are as in (7.2.1) and (7.2.2),

$$D_1 = \mathrm{diag}(1, e^{\frac{2\pi i}{n}}, \dots, e^{\frac{2\pi i}{n}(n-1)})$$

and

$$\hat{G} = \mathcal{F}_n G, \qquad \hat{B}^H = \mathcal{F}_m D B^H.$$

The optimal choice for θ is then $\theta = e^{\frac{\pi i}{N}}$, which ensures $\rho \geq 2\sin(\frac{\pi}{2N})$. Also compare the approach in [111].

7.3. Computing a tentative GCD

Let us assume for the time being that a tentative GCD degree k has been assigned. We will describe now a fast algorithm that either computes an ϵ-divisor, if it exists and can be reached by the algorithm, or else reports failure.

The first step involves the computation of cofactors. Recall from Section 6.3.3 that cofactors $p(x)$ and $q(x)$ can be computed from the following linear system:

$$
\begin{pmatrix}
u_0 & & v_0 & \\
u_1 & \ddots & v_1 & \ddots \\
\vdots & u_0 & \vdots & v_0 \\
\vdots & u_1 & \vdots & v_1 \\
u_n & \vdots & v_m & \vdots \\
& \ddots & \vdots & \ddots & \vdots \\
& u_n & & v_m
\end{pmatrix}
\begin{pmatrix}
q_0 \\
q_1 \\
\vdots \\
q_{m-k} \\
-p_0 \\
-p_1 \\
\vdots \\
-p_{n-k}
\end{pmatrix}
= 0. \qquad (7.3.1)
$$

A first approximation for an ϵ-GCD can then be retrieved as either $g(x) = u(x)/p(x)$ or $g(x) = v(x)/p(x)$.

Since the matrix in (7.3.1) – which is an $(n + m - k + 1) \times (n + m - 2k + 2)$ submatrix of the Sylvester matrix associated with $p(x)$ and $q(x)$ – is Toeplitz-like with displacement rank 2, the system can be solved in quadratic time by using the method described in Section 7.2.

Polynomial division is computed in our algorithms

- either by using the FFT (*i.e.* via evaluation/interpolation techniques, see [19] and Section B.2), which ensures good stability properties for small values of the given tolerance, together with a low computational cost,
- or by solving a linear system via GKO.

7.4. Fast iterative refinement

Next, we wish to apply iterative refinement to the approximate GCD and the cofactors obtained from (7.3.1). The discussion in Section 6.3.4 applies here as well. However, iterative refinement as described in Section 6.3.4 is an $\mathcal{O}(n^3)$ process, because at each iteration a linear least squares problem must be solved, and the most stable and reliable way to do this relies on the QR decomposition, which has a cubic cost, whereas we would like our algorithm to have a quadratic cost. Therefore we look for a way to apply fast factorization methods to iterative refinement, while retaining good convergence properties.

Recall from Section 6.3.4 that we wish to apply Newton's method to minimize the Euclidean norm of the function $F(\mathbf{z})$, where

$$
F(\mathbf{z}) =
\begin{bmatrix}
\mathcal{C}_p \mathbf{g} - \mathbf{u} \\
\mathcal{C}_q \mathbf{g} - \mathbf{v} \\
\|\mathbf{g}\|^2 - \|\mathbf{p}\|^2 - \|\mathbf{q}\|^2 - \tau
\end{bmatrix},
\qquad
\mathbf{z} =
\begin{bmatrix}
\mathbf{g} \\
\mathbf{p} \\
\mathbf{q}
\end{bmatrix}
$$

and τ is a constant. Observe that

$$
F : \mathbb{C}^M \longrightarrow \mathbb{C}^N
$$

where $M = m + n - k + 3$ and $N = m + n + 2$.

Let $\mathbf{z}_j$ be the Newton iterates and $\mathbf{z}^*$ be the limit point. Assume for the time being that $F(\mathbf{z}^*) = 0$, in which case Newton's method has a quadratic convergence rate.

At each step of the iterative method

$$
\mathbf{z}_{j+1} = \mathbf{z}_j - \eta_j
$$

the linear least squares problem

$$
J(\mathbf{z}_j)\eta_j = F(\mathbf{z}_j) \tag{7.4.1}
$$

must be solved. Recall that $J(\mathbf{z}_j)$ denotes the Jacobian matrix associated with the function F, computed at $\mathbf{z}_j$.

Remark 7.4.1. For every j, the matrix $J(\mathbf{z}_j)$ is Toeplitz-like and has displacement rank 5 with respect to the operator (2.2.5).

Proof. The assertion easily follows from the block Toeplitz structure of $J(\mathbf{z}_j)$ (see (6.3.19)). □

Set $\mathbf{x}_j = F(\mathbf{z}_j)$. The minimum norm solution of (7.4.1) is

$$
\eta_j = J(\mathbf{z}_j)^\dagger \mathbf{x}_j.
$$

In order to solve (7.4.1) with a quadratic computational cost, we would like to use the LU factorization, for which we have a fast algorithm, since $J(\mathbf{z}_j)$ is a displacement structured matrix. This is not admissible when seeking the solution of a general linear least squares system, but it is possible in this particular case, as it is proved below. We proceed as follows:

- Compute the factorization $J = LU$, where $J \in \mathbb{C}^{N \times M}$, $L \in \mathbb{C}^{N \times N}$ and $U \in \mathbb{C}^{N \times M}$. For the sake of simplicity, we are overlooking here the presence of permutation matrices due to the pivoting procedure; we can assume that either J or the vectors η_j and $\mathbf{x}_j = F(\mathbf{z}_j)$ have already undergone appropriate permutations.
  Consider the following block subdivision of the matrices L and U, where the left upper block has size $M \times M$:

$$L = \left[\begin{array}{c|c} L_1 & 0 \\ \hline L_2 & I \end{array}\right], \qquad U = \left[\begin{array}{c} U_1 \\ \hline 0 \end{array}\right].$$

Analogously, let $\mathbf{x}_j = \left[\begin{array}{c} \mathbf{x}_j^{(1)} \\ \hline \mathbf{x}_j^{(2)} \end{array}\right]$ and observe that $L^{-1} = \left[\begin{array}{c|c} L_1^{-1} & 0 \\ \hline -L_2 L_1^{-1} & I \end{array}\right].$

- Let $\mathbf{y}_j = L_1^{-1}\mathbf{x}_j^{(1)}$. If U_1 is nonsingular, then compute $\mathbf{w}_j$ as solution of $U_1 \mathbf{w}_j = \mathbf{y}_j$. Else, consider the block subdivision

$$U_1 = \left[\begin{array}{c|c} U_{11} & U_{12} \\ \hline 0 & 0 \end{array}\right], \qquad \mathbf{w}_j = \left[\begin{array}{c} \mathbf{w}_j^{(1)} \\ \hline \mathbf{w}_j^{(2)} \end{array}\right], \qquad \mathbf{y}_j = \left[\begin{array}{c} \mathbf{y}_j^{(1)} \\ \hline \mathbf{y}_j^{(2)} \end{array}\right],$$

such that U_{11} is nonsingular; set all the entries of $\mathbf{w}_j^{(2)}$ equal to zero, and compute $\mathbf{w}_j^{(1)}$ as solution of $U_{11}\mathbf{w}_j^{(1)} = \mathbf{y}_j^{(1)}$.
- If J is nearly rank deficient (as detected from the fast LU factorization), find a basis for $\mathcal{K} = \ker J$.
- Subtract from $\mathbf{w}_j$ its projection on $\mathcal{K}$, thus obtaining a vector χ_j. This is the vector that will be used as approximation of a solution of the linear least squares system in the iterative refinement process.

Let us now examine how χ_j is different from the "true" solution η_j of (7.4.1). Let $\mathcal{R}$ be the subspace of $\mathbb{C}^N$ spanned by the columns of J. We have

$$\mathbb{C}^N = \mathcal{R} \oplus \mathcal{R}^\perp. \tag{7.4.2}$$

Let

$$\mathbf{x}_j = \alpha_j + \beta_j$$

be the decomposition of $\mathbf{x}_j$ with respect to (7.4.2), i.e. we have $\alpha_j \in \mathcal{R}$ and $\beta_j \in \mathcal{R}^\perp$).

Notice that the Moore-Penrose pseudoinverse of J acts on $\mathbf{x}_j$ as follows: $J^\dagger \alpha_j$ is the preimage of α_j with respect to J and it is orthogonal to $\mathcal{K}$, whereas $J^\dagger \beta_j$ is equal to zero.

The LU-based procedure, on the other hand, acts exactly like $J^\dagger$ on α_j, whereas the component β_j is not necessarily sent to 0. Therefore,

χ_j is the sum of η_j and of the preimage of β_j with respect to the LU decomposition. If the problem is not ill-conditioned, it is reasonable to assume that the norm of this additional component is roughly of the order $\|\beta_j\|_2$.

In a general linear least squares problem, there is no reason for $\|\beta_j\|_2$ to be significantly smaller than $\|\mathbf{x}_j\|_2$. In our case, though, the Taylor expansion of $F(\mathbf{z})$ yields:

$$0 = F(\mathbf{z}^*) = F(\mathbf{z}_j) - J(\mathbf{z}_j)\epsilon_j + \mathcal{O}(\|\epsilon_j\|_2^2), \qquad (7.4.3)$$

where $\epsilon_j = \mathbf{z}_j - \mathbf{z}^*$. It follows from (7.4.3) that

$$\mathbf{x}_j = J(\mathbf{z}_j)\epsilon_j + \mathcal{O}(\|\epsilon_j\|_2^2).$$

Since $J(\mathbf{z}_j)\epsilon_j \in \mathcal{R}$, we conclude that $\|\beta_j\|_2 = \mathcal{O}(\|\epsilon_j\|_2^2)$. Therefore, we have some theoretical basis, together with heuristic evidence, showing that Newton's method applied to the iterative refinement of the polynomial GCD preserves its quadratic convergence rate, even though the linear least squares problems (7.4.1) are solved via the LU factorization of the Jacobian. Of course it should be kept in mind that the caveat of Section 6.3.5, concerning the convergence of this iterative process and the number of iterations it requires, still holds.

Example 7.4.2. This example shows the practical quadratic convergence of the iterative refinement. Let $p(x)$, $q(x)$ and $g(x)$ be polynomials of degree 10 whose coefficients are random real numbers in the interval $[-5, 5]$ and define $u(x) = p(x)g(x)$, $v(x) = q(x)g(x)$. The fast method presented in the previous sections is then applied to compute a tentative GCD and cofactors, to which we add dense perturbations of the order of 0.1. Fast iterative refinement, as described above, is then applied. The following table shows the behaviour of $\|\epsilon_n\|$ at each iteration:

n	$\|\epsilon_n\|$
1	1.32×10^{-1}
2	2.02×10^{-3}
3	1.74×10^{-6}
4	5.67×10^{-12}
5	0

7.4.1. Iterative refinement with line search

The iterative process ends when at least one of the following criteria is satisfied:

1. the residual (that is, the Euclidean norm of the function $F(\mathbf{z})$) becomes smaller that a fixed threshold,
2. the number of iterations reaches a fixed maximum,
3. the residual given by the last iteration is greater that the residual given by the previous iteration.

The purpose of the third criterion is to avoid spending computational effort on tentative GCDs that are not in fact suitable candidates. However, its use with Newton's method may pose some difficulties, because it is generally difficult to predict the global behaviour of this method; in particular, it might happen that the residual does not decrease monotonically. The usual way to overcome this obstacle is to use instead a relaxed version of Newton that includes a line search. More precisely, instead of the iteration (6.3.20) one computes

$$\mathbf{z}_{j+1} = \mathbf{z}_j - \alpha_j \tilde{J}(\mathbf{z}_j)^\dagger \tilde{F}(\mathbf{z}_j), \qquad (7.4.4)$$

where α_j is chosen – using a one-dimensional minimization method – so as to approximately minimize the norm of $\tilde{F}(\mathbf{z}_j)$.

The drawback of this technique is that it slows down convergence: the quadratic convergence that was one of the main interesting points of Newton's method is lost if one consistently performs iterations of the type (7.4.4). This motivates the use of a hybrid method: At each step, the algorithm evaluates the descent direction $\tilde{J}(\mathbf{z}_j)^\dagger \tilde{F}(\mathbf{z}_j)$ and checks if a pure Newton step (that is, (7.4.4) with $\alpha_j = 1$) decreases the residual. If this is the case, then the pure Newton step is actually performed; otherwise, α_j and subsequently $\mathbf{z}_{j+1}$ are computed by calling a line search routine. In this way, most of the optimization work is still performed by pure Newton iterations, so that the overall method remains computationally cheap; the line search, called only when necessary, is helpful in some difficult cases and ensures that the method has a sound theoretical basis. See [13] for a discussion and implementation of this variant. Some numerical results are reported in Section 8.1, Example 8.1.1.

7.5. Choice of a tentative degree

So far, we have developed a method to determine whether two given polynomials $u(x)$ and $v(x)$ have an ϵ-divisor of given degree k. We must now face the problem of choosing suitable values for k, so as to find an ϵ-divisor of maximum degree without having to run the "ϵ-divisor" test too many times.

A first option is to apply a bisection technique:

- Is there an ϵ-divisor of degree $k = m/2$?
- If the answer is yes, try again with $k = 3/4m$;

- If the answer is no, try again with $k = 1/4m$,
- and so on until the method converges to the ϵ-GCD degree.

Bisection requires to test the existence of an approximate divisor $\log_2 n$ times and therefore preserves the overall quadratic cost of the method.

In order to determine a tentative ϵ-GCD degree, however, one may also exploit the properties of the Sylvester or Bézout matrices associated with $u(x)$ and $v(x)$.

It is a well known fact that the dimension of the null space of either matrix is equal to the degree of the exact GCD of $u(x)$ and $v(x)$. Similarly, finding the ϵ-GCD degree is a matter of approximate rank determination, and fast LU factorization methods might provide reasonably useful values for a tentative ϵ-GCD degree.

7.5.1. Is LU factorization rank-revealing?

In exact arithmetic, an LU factorization is considered rank revealing if it assumes the form (see [70])

$$\Gamma_1 A P_2 = \begin{bmatrix} L_{11} & 0 \\ L_{12} & I \end{bmatrix} \begin{bmatrix} U_{11} & U_{12} \\ 0 & 0 \end{bmatrix},$$

with block sizes $\begin{bmatrix} (n-r) \times (n-r) & (n-r) \times r \\ r \times (n-r) & r \times r \end{bmatrix}$,

where A is any $n \times n$ matrix with null space of dimension r, L_{11} and U_{11} are nonsingular, and P_1, P_2 are permutation matrices depending on the chosen pivoting strategy. Gaussian elimination with no or partial pivoting does not, in general, yield a rank-revealing LU factorization, whereas rook, almost complete or complete pivoting do.

However, the relationship with the notion of an (approximate) rank-revealing factorization in a numerical setting is not straightforward. For instance, one would expect that, if A is nearly rank deficient, then one or more small pivots are found in U. But this might not be true, as shown by the following counterexample (Wilkinson):

$$W = \begin{pmatrix} 1 & -1 & \cdots & \cdots & -1 \\ 0 & 1 & -1 & \cdots & -1 \\ \vdots & \ddots & \ddots & \ddots & \vdots \\ \vdots & & \ddots & 1 & -1 \\ 0 & \cdots & \cdots & 0 & 1 \end{pmatrix} \in \mathbb{R}^{n \times n}.$$

If n is large enough, then W has a small singular value; but, if we apply GECP, the pivots in the (trivial) LU factorization of W are all equal to 1.

We face similar difficulties in the case of fast LU with almost complete pivoting. Still, something can be said on the relationship between fast LU factorization and approximate rank.

Assume that a small pivot a is found as the fast Gaussian elimination with almost complete pivoting of $C \in \mathbb{C}^{n \times n}$ is carried out, and denote with $\tilde{U} \in \mathbb{C}^{k \times k}$ the right lower block of the matrix, which has not been yet factorized. Then it follows from Lemma 7.1.1 that

$$|a| \leq \|\tilde{U}\|_{\max} \leq \sqrt{k}\rho|a|$$

and therefore, using well known matrix norm inequalities:

$$|a| \leq \|\tilde{U}\|_2 \leq k^{3/2}\rho|a|. \tag{7.5.1}$$

Remark 7.5.1. Let σ_{n-k} be the $n - k$-th singular value of C. Then it follows from (7.5.1) that

$$\sigma_{n-k} \leq \|\tilde{U}\|_2 \leq k^{3/2}\rho|a|.$$

So, fast Gaussian elimination with almost complete pivoting can actually provide an upper bound for the approximate rank of a structured matrix.

Observe that the incomplete fast LU factorization computes a Cauchy-like perturbation matrix ΔC (obtained from $\tilde{U}$ by applying the computed permutations P_1 and P_2), such that $C - \Delta C$ has rank $n - k$.

Now, let $u_\epsilon(x)$ and $v_\epsilon(x)$ be polynomials of minimum norm and same degrees as $u(x)$ and $v(x)$, such that $u + u_\epsilon$ and $v + v_\epsilon$ have an exact GCD of degree k. Assume $\|u_\epsilon\|_2 \leq \epsilon$ and $\|v_\epsilon\|_2 \leq \epsilon$. Let C_ϵ be the Cauchy-like matrix obtained via Proposition 2.4.2 from the Sylvester matrix S_ϵ associated with $u_\epsilon(x)$ and $v_\epsilon(x)$. Then $C + C_\epsilon$ has rank $n - k$, too.

If we assume that $\|\Delta C\|_2$ is very close to the minimum norm of a Cauchy-like perturbation that decreases the rank of C to $n - k$, then we have

$$|a| \leq \|\tilde{U}\|_2 = \|\Delta C\|_2 \leq \|C_\epsilon\|_2 = \|S_\epsilon\|_2 \leq \epsilon\sqrt{n+m}. \tag{7.5.2}$$

Therefore, if $|a| > \epsilon/\sqrt{n+m}$, then $u(x)$ and $v(x)$ cannot have an ϵ-divisor of degree k. This gives an upper bound on the ϵ-GCD degree based on the absolute values of the pivots found while applying the fast Gaussian elimination to C.

Of course, since the starting hypothesis might not be correct, this is only a heuristic criterion (which seems nevertheless to work quite well in practice). When it is applied, the GCD algorithm should check whether it actually provides an upper bound on the GCD degree.

7.6. The Fastgcd algorithm

The algorithm goes as follows.

Algorithm Fastgcd

Input: polynomials $u(x)$ and $v(x)$ and a tolerance ϵ.
Output: an ϵ-GCD $g(x)$; a backward error (residual of the GCD system); possibly perturbed polynomials $\hat{u}(x)$ and $\hat{v}(x)$ and cofactors $p(x)$ and $q(x)$.

- Compute the Sylvester matrix S associated with $u(x)$ and $v(x)$;
- Use Lemma 2.4.2 to turn S into a Cauchy-like matrix C;
- Perform fast Gaussian elimination with almost complete pivoting on C; stop when a pivot a such that $|a| < \epsilon/\sqrt{n+m}$ is found; let k_0 be the order of the not-yet-factored submatrix $\tilde{U}$ that has a as upper left entry;
- Choose $k = k_0$ as tentative GCD degree;
- Is there an ϵ-divisor of degree k?
- If yes, check for $k + 1$; if there is also an ϵ-divisor of degree $k + 1$, keep checking for increasing values of the degree until a maximum is reached (*i.e.* a degree is found for which there is no ϵ-divisor);
- If not, keep checking for decreasing values of the degree, until an ϵ-divisor (and GCD) is found.

Of course, when implementing the algorithm, it is not necessary to write explicitly S, C and the other matrices that are to be factorized; it is cheaper to use the displacement generators throughout the process.

It should also be pointed out that the algorithm generally outputs an approximate GCD with complex coefficients, even if $u(x)$ and $v(x)$ are real polynomials. This usually allows for a higher GCD degree or a smaller backward error. The drawback is that it would not be possible to compute an ϵ-GCD with real coefficients, if one were required.

Observe that a slightly different version of the above algorithm is still valid by replacing the Sylvester matrix with the Bezoutian. A tentative GCD degree is determined as in the Sylvester case (recall that the Bézout matrix is Toeplitz-like), whereas the coefficients are computed applying Theorem 2.6.13. Roughly speaking, in this modified version the size of the problem is reduced by a factor of 2, with clear computational advantage. However, experimental data show that, from the point of view of numerical stability, using Theorem 2.6.13 might sometimes be less advantageous than solving the Sylvester-like system.

Chapter 8
Numerical tests

The algorithms presented in Chapters 6 and 7 have been implemented in Matlab and applied to a wide variety of test polynomials, chosen as to be representative of the main difficulties that are typical of approximate GCD algorithms.

For each pair of polynomials, we show the results given by algorithms TdBez, PivQr and Fastgcd, and, for comparison purpose, the results given by other methods such as UVGCD by Zhonggang Zeng ([144]), in its version for Matlab, QRGCD by Corless *et al.* ([41]), and the STLN method developed by Kaltofen *et al.* ([81]). Implementation of the latter two methods for Maple are available and have been applied with precision fixed to 16 digits.

It must be pointed out that comparison with the STLN method is not straightforward, since this methods follows an optimization approach, *i.e.*, it takes two (or more) polynomials and the desired GCD degree k as input, and seeks a perturbation of minimum norm such that the perturbed polynomials have an exact GCD of degree k. Moreover, the algorithms UVGCD and STLN do not normalize the input polynomials, whereas QRGCD and Fastgcd do; therefore all test polynomials are normalized (with unitary Euclidean norm) beforehand.

The accuracy of the results has been evaluated by taking into account the following parameters:

deg: the degree of the computed approximate GCD. This result is particularly interesting when the degree of the ϵ-GCD of the test polynomials is very sensitive to the choice of ϵ;

res: the Euclidean norm of the residual associated with the GCD system, *i.e.*, the quantity $\sqrt{\|u - \hat{u}\|_2^2 + \|v - \hat{v}\|_2^2}$, where $u(x)$ and $v(x)$ are the input polynomials and $\hat{u}(x)$ and $\hat{v}(x)$ are the perturbed polynomials having an exact GCD. Observe that it must be res $\leq \sqrt{2}\epsilon$;

cwe: the coefficient-wise error on the computed ϵ-GCD. This parameter is meaningful when the test polynomials have an exact GCD, which is known beforehand; the computed ϵ-GCD is then expected to have the "correct" degree and to be close to the exact GCD. More precisely, if $g(x) = \sum_{j=0}^{k} g_j x^j$ is the exact GCD and $\hat{g}(x) = \sum_{j=0}^{k} \hat{g}_j x^j$ is the computed ϵ-GCD, then the coefficient-wise error is defined as

$$\text{cwe} = \max \frac{|g_j - \hat{g}_j|}{|g_j|},$$

where the maximum is taken over all indices $j = 0, 1, \ldots, k$ such that $g_j \neq 0$.

The tests that involve comparisons with UVGCD have been performed using Matlab 6.1 on an Intel Pentium 4, 2.66 GHz, 504 MB RAM, with operative system Microsoft Windows XP Professional. All the other tests have been carried out using Matlab 7.1 in a Unix environment, on a machine equipped with two Intel Xeon processors and 6 GB RAM.

8.1. Tolerance-sensitive degree

The first example in this section is taken from [144].

Example 8.1.1. Let

$$u(x) = \prod_{1}^{10} (x - x_j), \qquad v(x) = \prod_{1}^{10} (x - x_j + 10^{-j}),$$

with $x_j = (-1)^j (j/2)$.

The roots of $u(x)$ and $v(x)$ have decreasing distances $0.1, 0, 01, 0.001$, etc. The degree of the ϵ-GCD of $u(x)$ and $v(x)$ is therefore expected to be very sensitive with respect to the choice of ϵ. This example is also expected to be ill-conditioned.

We show here the ϵ-GCD degree and the residual found for different values of ϵ. A blank line means that the obtained result is the same as in the line above. Notice that for this experiment we also show the results obtained using the algorithm Fastgcd with line search, as described in Section 7.4.1.

ϵ	TdBez		PivQr	
	deg	res	deg	res
10^{-2}	9	0.004	9	0.004
10^{-3}	8	1.73×10^{-4}	8	1.73×10^{-4}
10^{-4}	7	7.09×10^{-6}	7	7.09×10^{-6}
10^{-5}				
10^{-6}	6	1.83×10^{-7}	5	4.49×10^{-9}
10^{-7}	5	4.49×10^{-9}		
10^{-8}				
10^{-9}	4	8.40×10^{-11}	2	2.25×10^{-14}

ϵ	Fastgcd w.l.s.		Fastgcd		UVGCD	
	deg	res	deg	res	deg	res
10^{-2}	9	0.0045	9	0.0045	9	0.0041
10^{-3}	8	2.63×10^{-4}	8	2.63×10^{-4}	8	1.73×10^{-4}
10^{-4}	7	9.73×10^{-6}	7	9.73×10^{-6}	8	1.73×10^{-4} (*)
10^{-5}					4	1.80×10^{-5}
10^{-6}	6	2.78×10^{-7}	4	6.26×10^{-10}	2	2.25×10^{-14}
10^{-7}	5	8.59×10^{-9}				
10^{-8}						
10^{-9}	4	6.26×10^{-10}				

(*) Here UVGCD has been called with tolerance 10^{-4} but outputs too high a degree, along with a residual that exceeds the given tolerance, probably due to the different definition of residual employed in the termination criteria of the program.

The results show that our algorithms are able – to slightly different extents – to separate the approximate GCD degrees while the tolerance decreases. In particular, it can be seen that incorporating a line search has beneficial effects in this particularly sensitive example.

The algorithm QRGCD outputs failure for all the values of ϵ smaller than 10^{-2}. The following table shows the residuals computed by STLN for several values of the degree.

deg (GCD)	res
9	5.65×10^{-3}
8	2.44×10^{-4}
7	1.00×10^{-5}
6	2.58×10^{-7}
5	6.34×10^{-9}
4	1.20×10^{-10}

The roots of the test polynomials in Example 8.1.1 are all real. We propose now a two-dimensional version of the example, in which the roots
have nonzero imaginary parts.

Example 8.1.2. Let $\{x_j\}_{j=1,\ldots16}$ be the complex numbers having real
parts equal to ±0.35 or ±1.05 and imaginary parts equal to ±0.35 or
±1.05, as in Figure 8.1. Define $y_j = x_j + 10^{-j}(1 + \hat{\imath})$ for every $j =
1, \ldots 16$. Let

$$u(x) = \prod_{j=1}^{16}(x - x_j), \qquad v(x) = \prod_{j=1}^{16}(x - y_j).$$

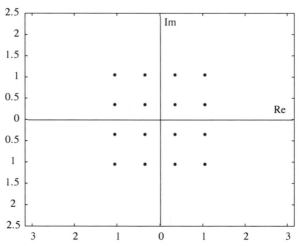

Figure 8.1. Roots $\{x_j\}_{j=1,\ldots16}$ of the polynomial $u(x)$ in Example 8.1.2.

Again, we expect the ϵ-GCD degree to be very sensitive to the choice of
ϵ. The tables show ϵ-GCD degree and residual for several values of the
tolerance.

ϵ	TdBez		PivQr	
	deg	res	deg	res
10^{-2}	15	0.0121	1	2.18×10^{-5}
10^{-3}	6	1.13×10^{-6}		
10^{-4}				
10^{-5}			ϵ-coprime	
10^{-6}				
10^{-7}	3	7.36×10^{-8}		
10^{-8}	ϵ-coprime			

ϵ	Fastgcd		UVGCD	
	deg	res	deg	res
10^{-2}	14	0.0016	13	0.0285 $^{(**)}$
10^{-3}	13	5.42×10^{-5}	7	0.0034 $^{(**)}$
10^{-4}				ϵ-coprime
10^{-5}	11	2.56×10^{-6}		
10^{-6}	10	1.47×10^{-7}		
10^{-7}	9	1.92×10^{-8}		
10^{-8}	8	3.38×10^{-9}		
10^{-9}		ϵ-coprime		

$^{(**)}$ Here UVGCD outputs a residual that exceeds the given tolerance, probably due to the different definition of residual employed in the termination criteria of the program.

The algorithm Fastgcd clearly performs quite well. Because the ϵ-GCDs have possibly non conjugate complex roots, it is to be expected that the algorithm TdBez might output approximate GCDs of low degree. Still, it gives better results than UVGCD.

8.2. Mignotte-like polynomials

We present here some examples where the pair of input polynomials is given by a Mignotte-like polynomial and its derivative.

Example 8.2.1. Let

$$u(x) = x^{20} + (x - 1/5)^7, \qquad v(x) = u'(x).$$

The polynomials $u(x)$ and $v(x)$ are coprime, but they have a nontrivial ϵ-GCD for ϵ large enough, due to the presence of a cluster of roots.

The following table shows the degrees of the ϵ-GCD found by Fastgcd and UVGCD with the polynomials of Example 8.2.1, for several values of ϵ.

ϵ	deg (Fastgcd)	ϵ	deg (UVGCD)
$> 10^{-7}$	6	$> 10^{-8}$	6
$10^{-8} \ldots 10^{-9}$	4	10^{-8}	ϵ-coprime
10^{-10}	3		
10^{-11}	ϵ-coprime		

We next show an example of higher degree.

Example 8.2.2. Let

$$u(x) = x^{100} + (x - 1/2)^{17}, \qquad v(x) = u'(x).$$

As in Example 8.2.1, the polynomials $u(x)$ and $v(x)$ are coprime, but they have a nontrivial ϵ-GCD for ϵ large enough.

The degrees of the ϵ-GCD found by Fastgcd and UVGCD with the polynomials of Example 8.2.2 are shown in the following table.

ϵ	deg (Fastgcd)	ϵ	deg (UVGCD)
10^{-1}	99	$10^{-1} \ldots 10^{-2}$	99
$10^{-2} \ldots 10^{-5}$	17	$10^{-3} \ldots 10^{-5}$	17
$10^{-6} \ldots 10^{-7}$	6	$10^{-6} \ldots 10^{-11}$	16
$10^{-8} \ldots 10^{-9}$	4	10^{-12}	ϵ-coprime
10^{-10}	3		
10^{-11}	ϵ-coprime		

Also see Section 8.10 for more experimental results on the approximate GCD of Mignotte-like polynomials.

8.3. High GCD degree

This example, taken from [144], uses polynomials such that their GCD has a large degree.

Example 8.3.1. Let $u_n(x) = g_n(x)p(x)$ and $v_n(x) = g_n(x)q(x)$, where $q(x) = \sum_{j=0}^{3} x^j$ and $p(x) = \sum_{j=0}^{4} (-x)^j$ are fixed polynomials and $g_n(x)$ is a polynomial of degree n whose coefficients are random integer numbers in the range $[-5, 5]$.

The following table shows the residuals and the coefficient-wise errors on the computed GCD for large values of n. All the algorithms give very accurate results and do not encounter difficulties related to the high degree of the polynomials. The STLN technique is also successful (results are not reported here). QRGCD, on the contrary, fails for degree 100 or larger, perhaps due to the fact that the common roots of these test polynomials are placed near the unitary circle.

n	TdBez		PivQr	
	res	cwe	res	cwe
50	1.94×10^{-16}	8.33×10^{-16}	1.71×10^{-16}	9.71×10^{-16}
100	2.27×10^{-16}	8.88×10^{-16}	1.96×10^{-16}	8.88×10^{-16}
200	1.97×10^{-16}	1.67×10^{-15}	1.74×10^{-16}	9.71×10^{-16}
500	1.90×10^{-16}	1.11×10^{-15}	1.80×10^{-16}	7.11×10^{-15}
1000	1.81×10^{-16}	1.33×10^{-15}	1.74×10^{-16}	1.78×10^{-15}

n	Fastgcd		UVGCD	
	res	cwe	res	cwe
50	2.51×10^{-16}	9.82×10^{-16}	1.77×10^{-16}	8.88×10^{-16}
100	2.65×10^{-16}	1.04×10^{-16}	1.83×10^{-16}	6.66×10^{-16}
200	2.67×10^{-16}	1.30×10^{-15}	1.61×10^{-16}	9.71×10^{-16}
500	3.60×10^{-16}	2.87×10^{-15}	1.96×10^{-16}	1.11×10^{-15}
1000	2.0×10^{-16}	2.62×10^{-15}	1.85×10^{-16}	1.22×10^{-15}

8.4. An ill-conditioned case

The polynomials in this example are specifically chosen so that the GCD problem is badly conditioned. The example is taken from [144].

Example 8.4.1. Let n be an even positive integer and $k = n/2$; define $u_n = g_n p_n$ and $v_n = g_n q_n$, where

$$g_n = \prod_{j=1}^{k}[(x - r_1\alpha_j)^2 + r_1^2\beta_j^2], \qquad p_n = \prod_{j=1}^{k}[(x - r_2\alpha_j)^2 + r_2^2\beta_j^2],$$

$$q_n = \prod_{j=k+1}^{n}[(x - r_1\alpha_j)^2 + r_1^2\beta_j^2], \qquad \alpha_j = \cos\frac{j\pi}{n}, \qquad \beta_j = \sin\frac{j\pi}{n},$$

for $r_1 = 0.5$ and $r_2 = 1.5$. The roots of u_n and v_n lie on the circles of radius r_1 and r_2.

The following tables show the errors given by the examined GCD methods as n increases. All the algorithm give quite accurate results in spite of the ill conditioning, except for the case $n = 20$.

n	TdBez		PivQr	
	res	cwe	res	cwe
12	6.29×10^{-15}	9.14×10^{-12}	9.87×10^{-15}	5.19×10^{-12}
14	5.14×10^{-14}	5.72×10^{-12}	4.24×10^{-14}	3.62×10^{-11}
16	1.36×10^{-13}	1.06×10^{-10}	9.55×10^{-14}	3.48×10^{-10}
18	1.47×10^{-5}	3.20×10^{-8}	5.84×10^{-9}	4.53×10^{-8}
20	2.47×10^{-8}	0.003	fails $^{(*)}$	

n	Fastgcd		UVGCD	
	res	cwe	res	cwe
12	1.65×10^{-14}	4.05×10^{-12}	9.99×10^{-15}	2.75×10^{-12}
14	4.81×10^{-14}	2.06×10^{-11}	3.66×10^{-14}	6.60×10^{-11}
16	2.27×10^{-13}	8.55×10^{-10}	1.54×10^{-13}	3.69×10^{-10}
18	1.08×10^{-12}	5.50×10^{-9}	5.21×10^{-13}	1.17×10^{-8}
20	fails $^{(*)}$		1.59×10^{-12}	1.55×10^{-8}

(*) Here the algorithm does not recognize the "correct" approximate GCD degree.

8.5. Unbalanced coefficients

This example is also taken from [144]; its difficulty lies in the great variation in magnitude of the GCD coefficients.

Example 8.5.1. Let $u(x) = g(x)p(x)$ and $v(x) = g(x)q(x)$, where $p(x)$ and $q(x)$ are as in Example 8.3.1 and

$$g(x) = \sum_{j=0}^{15} c_j 10^{e_j} x^j,$$

where c_j and e_j are random integers in $[-5, 5]$ and $[0, 6]$ respectively.

In this example $g(x)$ is the GCD of $u(x)$ and $v(x)$ and the magnitude of its coefficients varies between 0 and 5×10^6. If an approximate GCD algorithm is applied and the coefficient-wise relative error θ is calculated, then $N = \log_{10} \theta$ is roughly the minimum number of correct digits for the coefficients of $g(x)$ given by the chosen method. 100 repetitions of this test are performed.

The results – which are quite accurate for all algorithms – are plotted in Figures 8.2, 8.3, 8.4 and 8.5. The average of the correct digits found by each algorithm is 10.88 for TdBez, 10.92 for PivQr, 10.60 for Fastgcd and 11.39 for UVGCD.

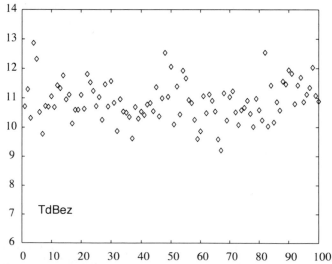

Figure 8.2. Number of correct digits found by the algorithm TdBez in Example 8.5.1

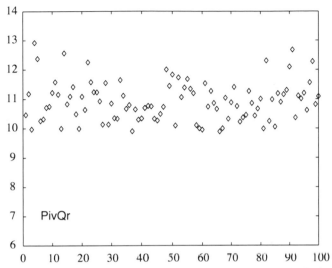

Figure 8.3. Number of correct digits found by the algorithm PivQr in Example 8.5.1

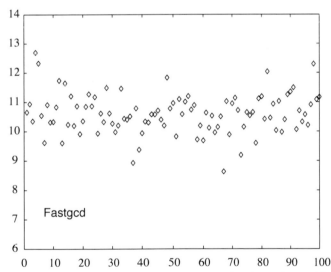

Figure 8.4. Number of correct digits found by the algorithm Fastgcd in Example 8.5.1

8.6. Multiple roots

In this section we compute the ϵ-GCD of polynomials with roots of high multiplicities and their derivatives.

Example 8.6.1. Let $u(x) = (x^3 + 3x - 1)(x - 1)^k$ for a positive integer k, and let $v(x) = u'(x)$. The GCD of $u(x)$ and $v(x)$ is $g(x) = (x - 1)^{k-1}$.

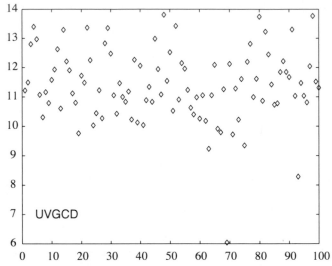

Figure 8.5. Number of correct digits found by the algorithm UVGCD in Example 8.5.1

The residuals computed for several values of k are shown here. The computed GCD degrees are understood to be correct. The algorithm TdBez has little difficulties with a moderate root multiplicity but fails for higher multiplicities, whereas PivQr finds residuals of the order of the machine precision.

k	TdBez	PivQr
25	4.48×10^{-16}	1.54×10^{-16}
35	fails $^{(*)}$	1.67×10^{-16}
45	fails $^{(*)}$	1.52×10^{-16}

$^{(*)}$ Here the algorithm does not recognize the "correct" approximate GCD degree.

The next table shows the coefficientwise errors found by Fastgcd and UVGCD; the residuals are in all cases of the same size as the machine precision.

k	Fastgcd	UVGCD
15	1.40×10^{-13}	3.84×10^{-13}
25	1.14×10^{-10}	3.61×10^{-12}
35	1.36×10^{-8}	1.03×10^{-9}
45	1.85×10^{-5}	1.72×10^{-9}

The algorithm UVGCD has been specifically designed for polynomials with multiple roots and is therefore very efficient. Fastgcd also provides good results, whereas QRGCD fails to find a gcd of correct degree as soon as the root multiplicity is larger than 25.

8.7. Small leading coefficient

A GCD with a small leading coefficient may represent in many cases a source of instability.

Example 8.7.1. For a given (small) parameter $\alpha \in \mathbb{R}$, let $g(x) = \alpha x^3 + 2x^2 - x + 5$, $p(x) = x^4 + 7x^2 - x + 1$ and $q(x) = x^3 - x^2 + 4x - 2$ and set $u(x) = g(x)p(x)$, $v(x) = g(x)q(x)$.

We applied Fastgcd and QRGCD to this example, with α ranging between 10^{-5} and 10^{-10}. It turns out that, for $\alpha < 10^{-5}$, QRGCD fails to recognize the correct gcd degree and outputs a gcd of degree 2. Fastgcd, on the contrary, always outputs a correct gcd, with a residual of 2.40×10^{-16}.

8.8. Effectiveness of the estimate on the ϵ-GCD degree

We recall here that in the implementation of the algorithm Fastgcd the approximate GCD degree is estimated using the heuristic criterion (7.5.2), which is usually expected to give an upper bound on the degree. It is interesting to describe the behaviour and effectiveness of this estimate on the numerical examples proposed in this chapter.

For examples which exhibit a clear gap in the magnitude of the pivots of the LU factorization of the associated Bézout or Sylvester matrix, the criterion (7.5.2) imediately gives the correct GCD degree. This is the case, for instance, when working with the test polynomials of Examples 8.3.1 or 8.5.1, and all the polynomials defined in Section 8.9. A typical situation is examined in Figure 8.6, which shows (in a logarithmic plot) the diagonal elements of the U factor in the fast LU factorization of the

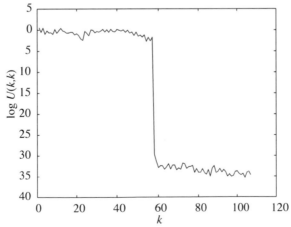

Figure 8.6. Example 8.3.1 with $n = 50$: behaviour of the pivots in the fast LU factorization of the associated Sylvester matrix.

Sylvester matrix associated with the polynomials of Example 8.3.1, with $n = 50$. Here we have $U(57, 57) = 0.1788$ and $U(58, 58) = 1.2788 \times 10^{-13}$, so it follows from (7.5.2) that, roughly speaking, any value of ϵ ranging from 10^{-12} to 1 yields the degree of the exact GCD.

In other, more difficult, cases (such as in Example 8.1.1), there is no clear gap in the magnitude of the pivots (see Figure 8.7), and indeed the choice of ϵ is crucial for the determination of the ϵ-GCD degree. The following table shows, for the polynomials of Example 8.1.1 and for several values of *epsilon*, the initial tentative degree given by (7.5.2) and the actual ϵ-GCD degree found by Fastgcd.

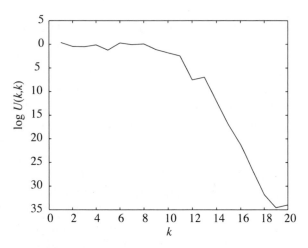

Figure 8.7. Example 8.1.1: behaviour of the pivots in the fast LU factorization of the associated Sylvester matrix

ϵ	initial degree	final degree
10^{-2}	9	9
10^{-3}	9	7
10^{-4}	7	7
10^{-5}	7	7
10^{-6}	6	6
10^{-7}	6	5
10^{-8}	6	5
10^{-9}	5	4

These results show that (7.5.2) provides an upper bound on the degree which is usually quite close to the actual ϵ-GCD degree.

8.9. Computation time

The crucial feature of the algorithm Fastgcd is its low computational cost, which a theoretical analysis shows to be quadratic in the degree of the input polynomials. Here this property is verified experimentally.

In the next two examples, we study how the running time (computed using the Matlab commands `tic` and `toc`) grows with the degree of the input polynomials. We define several sets of pairs of polynomials, parametrized by their degree N; for each set and for several increasing values of N we show a plot of the running time versus the degree in log-log scale (diamonds), along with a linear fit and its equation $y = \alpha x + \beta$. The experimental data then suggests that the running time of the algorithm should grow roughly as $\mathcal{O}(N^\alpha)$. The values obtained for α in these tests are always close to (and usually less than) 2.

Since the running time of a GCD algorithm may depend not only on the degrees of the input polynomials, but also on the degree of the GCD, we take care to provide examples where the GCD degree is almost as large as N, or about $N/2$, or very low. We point out that in all these tests the correct GCD is found with great accuracy (that is, with a residual of the order of the machine epsilon and with a coefficient-wise error on the computed GCD of about 10^{-15} or less).

The test polynomials for the first example are similar to the ones in Example 8.3.1. Since these polynomials have random coefficients, the running time is not strictly a function of the degree only, but also depends on the particular choice of the coefficients. As a consequence, the resulting growth rate for the running time may be biased. In order to overcome this difficulty, we repeat the process of choosing the polynomials and applying the algorithm Fastgcd ten times for each value of the degree, and take the average time as a result.

Example 8.9.1. 1. Let $u_n(x) = g(x)p_n(x)$ and $v_n(x) = g(x)q_n(x)$, where $p_n(x)$ and $q_n(x)$ are polynomials of degree n whose coefficients are random integer numbers in the range $[-5, 5]$, whereas the GCD $g(x) = \sum_{j=0}^4 (-1)^j x^j$ is a fixed polynomial of low degree. Here both $u_n(x)$ and $v_n(x)$ have degree $n + 4$. The results for n ranging between 100 and 350 are shown in Figure 8.8. Here we obtain $\alpha = 1.96$.

2. Let $u_n(x) = g_n(x)p_n(x)$ and $v_n(x) = g_n(x)q_n(x)$, where $g_n(x)$, $p_n(x)$ and $q_n(x)$ are polynomials of degree n whose coefficients are random integer numbers in the range $[-5, 5]$. Therefore both $u_n(x)$ and $v_n(x)$ have degree $2n$. Observe that in this example the GCD and the cofactors all have the same (large) degree. The results for n ranging between 25 and 250 are shown in Figure 8.9. Here we obtain $\alpha = 2.05$.

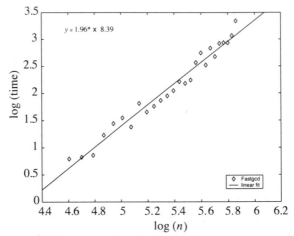

Figure 8.8. Running time for the algorithm Fastgcd in Example 8.9.1, with GCD of fixed small degree and cofactors of large degree.

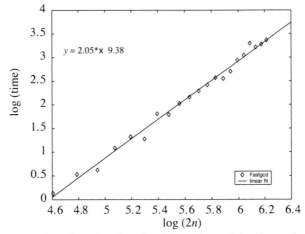

Figure 8.9. Running time for the algorithm Fastgcd in Example 8.9.1, with GCD and cofactors of equal degree.

3. Here $u_n(x)$ and $v_n(x)$ are taken exactly as in Example 8.3.1. The degrees of $u_n(x)$ and $v_n(x)$ are $n+4$ and $n+3$, respectively, whereas the GCD $g_n(x)$ has degree n. The results for n ranging between 100 and 350 are shown in Figure 8.10. Here we obtain $\alpha = 1.73$.

We perform now a similar test on polynomials whose coefficients are not randomly chosen and depend only on the degree. The polynomials are chosen so as to define well-conditioned problems, *i.e.*, the roots of each cofactor are well separated from the roots of the GCD and of the other cofactor. Here we simply consider the running time for each pair

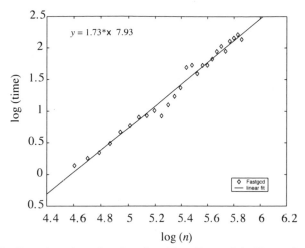

Figure 8.10. Running time for the algorithm Fastgcd in Example 8.9.1, with GCD of large degree and cofactors of fixed small degree.

of polynomials, without averaging on repeated trials. The degrees of the input polynomials range between 50 and 1000, roughly.

Example 8.9.2. Let k be a positive integer (which ranges between 1 and 20).

1. Let $n_1 = 25k$, $n_2 = 15k$, $n_3 = 10k$ and define $u_k(x) = g(x)p_k(x)$ and $v_k(x) = g(x)q_k(x)$, where

$$p_k(x) = (x^{n_1} - 1)(x^{n_2} - 2)(x^{n_3} - 3),$$
$$q_k(x) = (x^{n_1} + 1)(x^{n_2} + 5)(x^{n_3} + \hat{\imath}),$$
$$g(x) = x^4 + 10x^3 + x - 1.$$

Both cofactors have degree $50k$, whereas the GCD has a small fixed degree. The results are shown in Figure 8.11; we obtain $\alpha = 1.95$.

2. Let $n_1 = 10k$, $n_2 = 10k$, $n_3 = 5k$ and define $u_k(x) = g_k(x)p_k(x)$ and $v_k(x) = g_k(x)q_k(x)$, where

$$p_k(x) = (x^{n_1} + 10 - 0.7\hat{\imath})(x^{n_2} + 1 + 0.5\hat{\imath})(x^{n_3} + 0.01 + 0.3\hat{\imath}),$$
$$q_k(x) = (x^{n_1} - 11 - 2\hat{\imath})(x^{n_2} - 0.3 + 3\hat{\imath})(x^{n_3} + 5 - \hat{\imath}),$$
$$g_k(x) = (x^{n_1} - 1)(x^{n_2} - 2)(x^{n_3} - 3).$$

The cofactors and the GCD have degree $25k$. The results are shown in Figure 8.12; we obtain $\alpha = 1.96$.

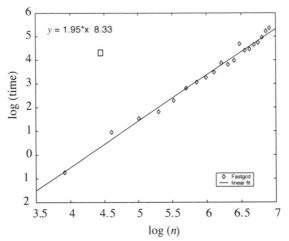

Figure 8.11. Running time for the algorithm Fastgcd in Example 8.9.2, with GCD of fixed small degree and cofactors of large degree.

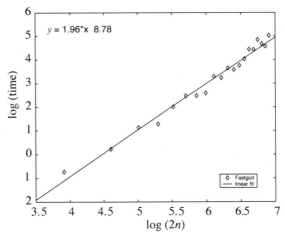

Figure 8.12. Running time for the algorithm Fastgcd in Example 8.9.2, with GCD and cofactors of equal large degrees.

3. Let $n_1 = 25k$, $n_2 = 15k$, $n_3 = 10k$ and define $u_k(x) = g_k(x)p(x)$ and $v_k(x) = g_k(x)q(x)$, where

$$p_k(x) = x^4 + (9 - 2\hat{\imath})x^3 + (29 - 14\hat{\imath})x^2 + (39 - 32)x + 18 - 24\hat{\imath},$$
$$q_k(x) = x^3 + (3 - 2\hat{\imath})x^2 - 10x - 24 + 18\hat{\imath},$$
$$g(x) = (x^{n_1} - 1)(x^{n_2} - 2)(x^{n_3} - 3).$$

The cofactors have fixed degrees 4 and 3, whereas the GCD has degree $50k$. The results are shown in Figure 8.13; we obtain $\alpha = 1.84$.

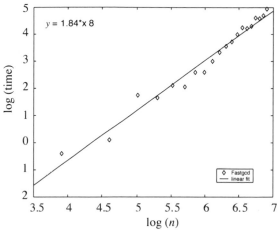

Figure 8.13. Running time for the algorithm Fastgcd in Example 8.9.2, with cofactors of fixed small degree and GCD of large degree.

As remarked above, the test polynomials in Examples 8.9.1 and 8.9.2 have been chosen so as to provide different combinations of degrees of GCDs and cofactors, in order to check whether Fastgcd slows down or speeds up under particular conditions. There do not seem to be remarkable differences in performance in the several tests proposed.

This is not the case for other methods; for example, UVGCD (whose running time has in general a cubic growth with the degree) is considerably faster when the cofactors have low degree, whereas it slows down remarkably when the cofactors have large degree.

We show now a comparison between the running times of the two methods. We point out here that several slightly different implementations of algorithm Fastgcd have been written. In particular, one implementation uses the Matlab built-in `qr` command to orthogonalize the displacement generators, whereas another one uses modified Gram-Schmidt orthogonalization. The first implementation has been used throughout all these numerical experiments.

The differences in performance between the two versions lie mainly in the required running time and depend on the Matlab version in use. The crucial point is the QR factorization (with rectangular Q and square R) of displacement generators. In Matlab 7 the built-in `qr` function is optimized for rectangular matrices, so it is preferable to use the implementation of Fastgcd that uses this command. On the contrary, in older versions of Matlab the Gram-Schmidt implementation is faster. Since comparisons with UVGCD have been carried out using Matlab 6.1, we have used the Gram-Schmidt version when comparing the running times.

We also show "extrapolated" running times for the qr-based version. These results are obtained as follows. First, for each polynomial pair the running times for the two versions of Fastgcd have been computed as the average on ten trials, using Matlab 7. Then, the running times obtained for the Gram-Schmidt version on Matlab 6 have been scaled according to the previous results. We emphasize that the results thus computed are not actual running times. However, they show how the QR-based implementation compares to the Gram-Schmidt one, and it is reasonable to assume that they closely resemble the performance of the qr-based implementation compared to UVGCD (which could not be directly verified).

The comparisons have been carried out on the polynomials of Example 8.9.2. Figure 8.14 shows the results for the first case (GCD of low fixed degree and cofactors of large, increasing degree), where UVGCD is particularly slow. Figure 8.14 refers to the case where the GCD and the cofactors have the same large degree. The results for the third case (cofactors of low fixed degree and GCD of large, increasing degree) are shown in Figure 8.16; this is the most favorable case for the algorithm UVGCD. All plots, however, clearly show how Fastgcd becomes faster for sufficiently high degrees.

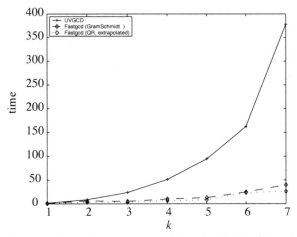

Figure 8.14. Running times for Example 8.9.2, with GCD of low degree.

8.10. A comparison with δ-GCD

This section is devoted to an experimental comparison between the "classical" coefficient-based definition of ϵ-GCD (*i.e.* Definition 1.1.3) and the root-based definition of δ-GCD given by Pan (see Section 1.4). It must be pointed out that in this type of comparison the use of a normalized definition of ϵ-GCD is crucial, because the definition of δ-GCD only

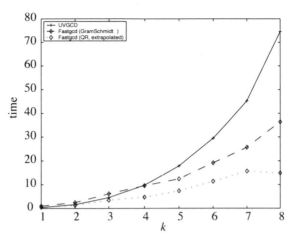

Figure 8.15. Running times for Example 8.9.2, with GCD and cofactors of equal degrees.

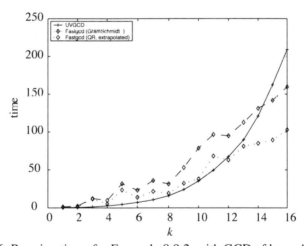

Figure 8.16. Running times for Example 8.9.2, with GCD of large degree.

involves roots and is therefore "blind" to any scaling of the coefficients. We recall that throughout this work polynomials with unitary Euclidean norm have been consistently used.

A comparison between the degrees of ϵ-GCDs and of δ-GCDs has been implicitly done in Section 8.1, where in each pair of test polynomials the roots have decreasing distances, defined *a priori*. Therefore it is easy to check how many pairs of roots are separated by a distance smaller than a fixed threshold δ; this number corresponds to the δ-GCD degree.

It is clear that the most interesting cases, where the degrees of ϵ-GCDs and of δ-GCDs behave in remarkably different ways, happen when multiple or clustered roots are present, rather than when the roots are well separated. For this reason we examine here the Mignotte-like polynomials defined in Examples 8.2.1 and 8.2.2.

The δ-GCD has been computed following the algorithm outlined in Section 1.4. The practical details are listed below:

- Given polynomials $u(x) = \sum_{i=0}^{n} u_i x^i$ and $v(x) = \sum_{i=0}^{m} v_i x^i$, compute their roots $\{\alpha_i\}_{i=1,\ldots,n}$ and $\{\beta_i\}_{i=1,\ldots,m}$. This has been done using MPSolve, a polynomial rootfinder designed and implemented by Bini and Fiorentino (see [14]), which can be downloaded at
  `http://www.dm.unipi.it/cluster-pages/mpsolve`.
  MPSolve relies on the GMP multiprecision package and finds polynomial roots with any prescribed accuracy, employing an adaptive design.
- Write the bipartite graph $G = (U \cup V, E)$, where $U = \{\alpha_i\}_{i=1,\ldots,n}$ and $V = \{\beta_i\}_{i=1,\ldots,m}$. Also write a matrix A associated to G; that is, A is an $n \times m$ matrix such that $A(i, j) \neq 0$ if and only if α_i and β_j are linked by an edge in G.
- Find a maximum matching by using the Matlab command `dmperm` applied to A.
- For each pair of roots (α_i, β_i), $i = 1, \ldots, k$ belonging to the maximum matching, set $w_i = (\alpha_i + \beta_i)/2$.
- Set $g(x) = \prod_{i=1}^{k}(x - w_i)$; this is the required δ-GCD.

The program MPSolve has also been used in [131] to compute a δ-GCD, and the resulting algorithm has undergone numerical tests and has been compared to methods for the computation of an ϵ-GCD. Here MPSolve has been used with the aim of finding polynomial roots with very high accuracy, so that the computed δ-GCD is very close to the theoretical one.

We list in the following table the degrees of the computed δ-GCDs of the polynomials $u(x)$ and $v(x)$ defined in Example 8.2.1 for several values of δ, along with the degrees of the ϵ-GCDs for comparison purposes. The ϵ-GCD degrees shown here have been chosen as the highest found by the examined ϵ-GCD methods. Figure 8.17 shows the roots of $u(x)$ and $v(x)$.

δ	deg(δ-GCD)	ϵ	deg(ϵ-GCD)
10^{-1}	19	10^{-6}	6
4×10^{-2}	12	10^{-8}	4
3×10^{-2}	8	10^{-10}	3
2×10^{-2}	6	10^{-11}	ϵ-coprime
10^{-3}	δ-coprime		

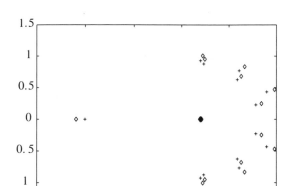

Figure 8.17. Roots of the polynomials $u(x)$ (diamonds) and $v(x)$ (crosses) defined in Example 8.2.1.

The following table shows the δ- and ϵ-GCD degrees for the polynomials in Example 8.2.2. The roots of these polynomials are plotted in Figure 8.18.

[h] δ	deg(δ-GCD)	ϵ	deg(ϵ-GCD)
10	99	10^{-2}	99
10^{-2}	32	10^{-5}	17
9×10^{-3}	24	10^{-6}	16
8×10^{-3}	18	10^{-12}	ϵ-coprime
7×10^{-3}	12		
6×10^{-3}	2		
3×10^{-3}	δ-coprime		

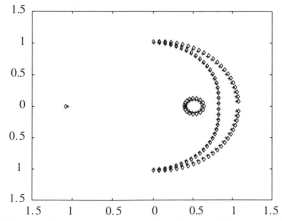

Figure 8.18. Roots of the polynomials $u(x)$ (diamonds) and $v(x)$ (crosses) defined in Example 8.2.2.

In the next example, we choose two polynomials with multiple roots, apply a perturbation of fixed magnitude on their coefficients and compute the δ-GCD degree of the resulting polynomials for several values of δ.

Example 8.10.1. Let

$$u(x) = \left(x - \frac{1}{2}\right)^{13}\left(x + \frac{1 + \hat{\imath}}{2}\right)^{11}\left(x + \frac{1 - \hat{\imath}}{2}\right)^{9},$$

$$v(x) = u'(x).$$

The GCD of $u(x)$ and $v(x)$ has a root of multiplicity 12 in $1/2$, a root of multiplicity 10 in $(-1 - \hat{\imath})/2$ and a root of multiplicity 8 in $(-1 + \hat{\imath})/2$.
   For a fixed $\epsilon > 0$, define

$$\hat{u}_\epsilon(x) = \frac{u(x)}{\|u\|_2} + u_\epsilon(x),$$

$$\hat{v}_\epsilon(x) = \frac{v(x)}{\|v\|_2} + v_\epsilon(x),$$

where $u_\epsilon(x)$ and $v_\epsilon(x)$ are perturbation polynomials of the same degrees as $u(x)$ and $v(x)$, such that $\|u_\epsilon\|_2 = \|v_\epsilon\|_2 = 1$. We have $\deg \epsilon$-$\mathrm{GCD}(\hat{u}_\epsilon, \hat{v}_\epsilon) =$
$\deg \mathrm{GCD}(u, v) = 31$.

   In the following tables, we choose some different values for ϵ (namely, 10^{-4}, 10^{-8}, 10^{-12} and 10^{-16}) and compute, for each of them, the degree of the δ-GCD of $\hat{u}_\epsilon(x)$ and $\hat{v}_\epsilon(x)$ for several values of δ. Figure 8.19

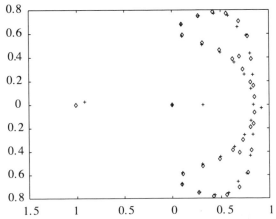

Figure 8.19. Roots of the polynomials $\hat{u}_\epsilon(x)$ (diamonds) and $\hat{v}_\epsilon(x)$ (crosses) defined in Example 8.10.1.

shows the roots of $\hat{u}_\epsilon(x)$ and $\hat{v}_\epsilon(x)$ for $\epsilon = 10^{-4}$.

$\epsilon = 10^{-4}$		$\epsilon = 10^{-8}$	
δ	deg(δ-GCD)	δ	deg(δ-GCD)
10^{-1}	31	10^{-1}	31
10^{-2}	12	10^{-2}	18
10^{-3}	δ-coprime	10^{-3}	δ-coprime

$\epsilon = 10^{-12}$		$\epsilon = 10^{-16}$	
δ	deg(δ-GCD)	δ	deg(δ-GCD)
10^{-1}	31	10^{-1}	31
10^{-2}	24	10^{-2}	28
10^{-3}	δ-coprime	10^{-3}	δ-coprime

Chapter 9
Generalizations and further work

The present work focuses on properties and computation of an approximate GCD of two univariate polynomials, which are expressed through their coefficients in the usual monomial basis.

There are several directions along which this problem can be generalized. One may want to compute an approximate GCD of more than two polynomials, or may need to deal with multivariate polynomials. For example, the application of polynomial GCD to image deblurring presented in [108] requires to compute an approximate GCD of bivariate polynomials.

Moreover, one may wish to define the input polynomials $u(x)$ and $v(x)$ in a different way than giving the coefficients in the monomial basis:

- $u(x)$ and $v(x)$ may be defined by their roots. A presentation of how the approximate GCD problem can be formulated and solved in this case has been given in Sections 1.4 and 1.5 (see also [103]); numerical experiments are found in Section 8.10;
- $u(x)$ and $v(x)$ may be defined by the values taken on a properly chosen set of points. This approach is pursued, in a wider context, in [3];
- $u(x)$ and $v(x)$ may be defined by their coefficients in a basis different from the monomial one. The GCD problem posed in this form is interesting from the point of view of structured matrices based methods because it leads to a reformulation of the resultant matrices involved.

Another line of research within the approximate GCD problem is the design of fast algorithms that exploit the structure properties of resultant matrices.

A fast algorithm based on fast LU decomposition has been described in Chapter 7. It would be interesting, however, to examine fast algorithms for the QR decomposition, since this type of matrix factorization is known to be closely related to polynomial GCD. Most efforts in this direction have so far produced unstable methods; the goal is therefore to achieve stability, too.

9.1. Approximate GCD of many polynomials

Methods for the exact computation of the GCD of a pair of polynomials are easily extended to many polynomials. If the GCD of $u_1(x), u_2(x), \ldots$ $\ldots, u_r(x)$ is desired, it suffices to compute

$$g_1(x) = u_1(x),$$
$$g_{j+1}(x) = \text{GCD}\,(g_j, u_{j+1}), \quad j = 1, \ldots, r - 1 \quad (9.1.1)$$

and $g_{r-1}(x)$ turns out to be the required GCD.

In the approximate case, though, the generalization is not so straightforward. First, we need to redefine the notion of approximate GCD. For instance, the generalization of Definition 1.1.3 (ϵ-GCD) goes as follows:

Definition 9.1.1. Given polynomials $u_1(x), u_2(x), \ldots, u_r(x)$ and a tolerance ϵ, the polynomial $g(x)$ is an ϵ-GCD of $u_1(x), u_2(x), \ldots, u_r(x)$ if:

  (i) $g(x)$ is an ϵ-divisor of $u_j(x)$ for all $j = 1, \ldots, r$;
  (ii) $g(x)$ has maximum degree among all polynomials that satisfy (i).

Computing the approximate equivalent of a sequence of polynomials $g_1(x), \ldots,$
$g_{r-1}(x)$ as in (9.1.1) yields in this case an ϵ-divisor of all the input polynomials, but possibly not an ϵ-GCD.

However, matrix-based methods for computing an approximate GCD of a pair of polynomials can be generalized to many polynomials if generalized resultant matrices are used. Indeed, some of the algorithms presented in Chapters 4 and 5 had actually been conceived in the framework of generalized resultants ([47, 81, 114, 144]).

A review of resultant-based techniques for approximate GCD computations with many polynomials, as well as a presentation of the generalized Sylvester matrix, is found in [83]; see also [34]. The so-called matrix pencil method ([82]) and subspace method ([109]) also rely on generalized resultants of Sylvester type. Moreover, Rupprecht's work ([114]) is based on the generalization of Sylvester subresultants.

The generalization of Bezoutians and companion matrix resultants is found in [5] and – in a different form and with applications to approximate polynomial GCD – in [46] and [47].

We recall here the definitions and main properties of generalized Sylvester and Bézout matrices.

9.1.1. Generalized resultants

Let $p(x), q^{(1)}(x), \ldots, q^{(h)}(x)$ be real or complex polynomials of degrees $n, m_1, \ldots, m_h$ respectively. Let $m = \max_{1 \le j \le h} m_j$ and assume that $n \le m$.

Define an $m \times (n + m)$ matrix associated with $p(x)$, as follows:

$$S_0 = \begin{bmatrix} p_n & p_{n-1} & p_{n-2} & \cdots & p_1 & p_0 & 0 & \cdots & 0 \\ 0 & p_n & p_{n-1} & p_{n-2} & \cdots & p_1 & p_0 & \ddots & \vdots \\ \vdots & \ddots & \ddots & \ddots & & & \ddots & \ddots & 0 \\ 0 & \cdots & 0 & p_n & p_{n-1} & p_{n-2} & \cdots & p_1 & p_0 \end{bmatrix}.$$

Analogously, define for each $j = 1, \ldots, h$ a matrix of size $n \times (n + m)$ associated with $q_j(x)$:

$$S_j = \begin{bmatrix} q_{m_j}^{(j)} & q_{m_j-1}^{(j)} & q_{m_j-2}^{(j)} & \cdots & q_1^{(j)} & q_0^{(j)} & 0 & \cdots & 0 \\ 0 & q_{m_j}^{(j)} & q_{m_j-1}^{(j)} & q_{m_j-2}^{(j)} & \cdots & q_1^{(j)} & q_0^{(j)} & \ddots & \vdots \\ \vdots & \ddots & \ddots & \ddots & & & \ddots & \ddots & 0 \\ 0 & \cdots & 0 & q_{m_j}^{(j)} & q_{m_j 1}^{(j)} & q_{m_j-2}^{(j)} & \cdots & q_1^{(j)} & q_0^{(j)} \end{bmatrix}.$$

Then the *generalized Sylvester matrix* associated with $p(x), q^{(1)}(x), \ldots, q^{(h)}(x)$ is defined as

$$S_G = \begin{bmatrix} S_0 \\ S_1 \\ \vdots \\ S_h \end{bmatrix}$$

and it is a matrix of size $(m + hn) \times (n + m)$.

The properties of S_G with respect to polynomial GCD are very similar to the properties of the Sylvester matrix defined for a pair of polynomials.

Theorem 9.1.2. *Let S_G be the generalized Sylvester matrix associated with polynomials $p(x), q^{(1)}(x), \ldots, q^{(h)}(x)$, defined as above. Then:*

(i) *the polynomials $p(x), q^{(1)}(x), \ldots, q^{(h)}(x)$ are coprime if and only if* rank $S_G = n + m$;

(ii) *if $\phi(x)$ is the GCD of $p(x), q^{(1)}(x), \ldots, q^{(h)}(x)$, then*

$$\text{rank } S_G = n + m - \deg \phi(x);$$

(iii) *if S_G is reduced to its row echelon form via elementary row operations, then the last nonzero row contains the coefficients of a (non monic) GCD.*

Let us now define generalized matrices of Bezoutian type. The *generalized companion matrix resultant* associated with polynomials $p(x)$, $q^{(1)}(x), \ldots, q^{(h)}(x)$ is defined as

$$R_{\mathcal{G}} = \begin{bmatrix} q^{(1)}(F_p) \\ q^{(2)}(F_p) \\ \vdots \\ q^{(h)}(F_p) \end{bmatrix},$$

where F_p is the Frobenius matrix of the polynomial $p(x)$, defined as in Section 2.7.

Analogously, the *generalized Bezoutian* is an $hn \times n$ matrix defined as

$$B_{\mathcal{G}} = \begin{bmatrix} \text{Bez}(p, q^{(1)}) \\ \text{Bez}(p, q^{(2)}) \\ \vdots \\ \text{Bez}(p, q^{(h)}) \end{bmatrix},$$

where $\text{Bez}(p, q^{(j)})$ is the ordinary Bezoutian associated with polynomials $p(x)$ and $q^{(h)}(x)$.

Theorem 9.1.3. *With the above notation, the following properties hold:*

(i) *the polynomials $p(x), q^{(1)}(x), \ldots, q^{(h)}(x)$ are coprime if and only if $B_{\mathcal{G}}$ has full rank;*
(ii) *if $\phi(x)$ is the GCD of $p(x), q^{(1)}(x), \ldots, q^{(h)}(x)$, then*

$$\deg\phi(x) = n - \text{rank } B_{\mathcal{G}};$$

(iii) *if $k = \deg\phi(x)$, and therefore rank $B_{\mathcal{G}} = n - k$, then the last $n - k$ columns of $B_{\mathcal{G}}$ are linearly independent;*
(iv) *if the columns of $B_{\mathcal{G}}$ are denoted as $\mathbf{c}_1, \ldots, \mathbf{c}_n$, then it follows from (iii) that each $\mathbf{c}_i$, with $1 \le i \le k$, can be written as a linear combination of $\mathbf{c}_{k+1}, \ldots, \mathbf{c}_n$:*

$$\mathbf{c}_{k-i} = \sum_{j=k+1}^{n} h_{k-i}^{(j)} \mathbf{c}_j, \qquad i = 0, \ldots, k - 1.$$

Then

$$\phi(x) = x^k + h_k^{(k+1)} x^{k-1} + h_{k-1}^{(k+1)} x^{k-2} + \cdots + h_2^{(k+1)} x + h_1^{(k+1)}$$

is the GCD of $p(x), q^{(1)}(x), \ldots, q^{(h)}(x)$.

It follows from the definitions and from Theorems 9.1.2 and 9.1.3 that many among the resultant matrix-based algorithms for the computation of the approximate GCD of a pair of polynomials can be extended to the case of many polynomials. Moreover, the generalized resultants presented in this section also display structure properties that may be exploited in the design of approximate GCD algorithms.

9.2. Approximate GCD of multivariate polynomials

The multivariate version of the approximate GCD problem is defined as in the univariate case, except that we expect to work with r-variate polynomials $u(x_1, \ldots, x_r)$ and $v(x_1, \ldots, x_r)$.

A study of the multivariate GCD problem from the point of view of the Euclidean algorithm had already been done in [25]. Generalizations of the Euclidean algorithm and polynomial remainder sequences are also used in [116] and [100], whereas Hensel lifting is employed in [148] and [147]. Moreover, a discussion of interpolation techniques applied to the approximate setting is found in [40].

Since we focus here on matrix-based methods, we point out that the notion of resultant matrix can be generalized to the multivariate case. Generalizations of the Sylvester matrix and subresultants, which we will describe now, are employed in [108, 57, 145, 81].

Given polynomials $u(x_1, \ldots, x_r)$ and $v(x_1, \ldots, x_r)$ of r-degrees n and m respectively, define the multivariate Sylvester matrix associated with u and v as

$$S(u, v) = \begin{bmatrix} C_m(u) & C_n(v) \end{bmatrix},$$

where $C_m(u)$ and $C_n(v)$ are multivariate convolution matrices (see Section B.1.2 for notation and definitions).

Moreover, we have (see [145]):

Proposition 9.2.1. *The GCD of $u(x_1, \ldots, x_r)$ and $v(x_1, \ldots, x_r)$ has r-degree k if and only if the matrix*

$$S_k(u, v) = \begin{bmatrix} C_{m-k}(u) & C_{n-k}(v) \end{bmatrix}$$

is rank-deficient with nullity one. If this is the case, then the null space of $S_k(u, v)$ is spanned by $[-\mathbf{q}^T \quad \mathbf{p}^T]^T$, where $\mathbf{q} \in \mathbb{C}^{\nu(m-k)}$ and $\mathbf{p} \in \mathbb{C}^{\nu(n-k)}$ are coefficient vectors of the cofactors.

A further generalization of the matrix $S_k(u, v)$ to the case of many polynomials is given by (see [81])

$$\tilde{S}_k(u, v) = \begin{bmatrix} C_{m_1-k}(u) & 0 & \cdots & 0 & C_{n-k}(v_1) \\ 0 & C_{m_2-k}(u) & & 0 & C_{n-k}(v_2) \\ \vdots & & \ddots & \vdots & \vdots \\ 0 & 0 & \cdots & C_{m_h-k}(u) & C_{n-k}(v_h) \end{bmatrix},$$

where $u, v_1, \ldots, v_h$ are r-variate polynomials, of r-degrees $n, m_1, \ldots, m_h$ respectively. Then we have:

Proposition 9.2.2. *The GCD of $u, v_1, \ldots, v_h$ has r-degree at least k if and only if $\tilde{S}_k(u, v)$ has rank deficiency at least one.*

It should be pointed out that convolution and Sylvester matrices for the multivariate case are often of very large size. It is therefore particularly important to exploit their sparsity and structure properties.

Similar generalizations can also be attempted for the Bézout matrix; see for instance [18] for a Bezoutian matrix associated with bivariate Bernstein polynomials.

9.3. Polynomial GCD in other bases

Most approaches to the computation of polynomial GCD assume that the input polynomials are expressed in the power basis. If they are expressed with respect to a different basis, it is certainly possible to tranform the polynomials into the power basis, compute a GCD and, if necessary, transform it back to the original basis. This, however, might turn out to be an ill conditioned procedure.

Assuming that we plan to apply a resultant matrix-based method, another – and often preferable – possibility is to avoid changes of basis and employ resultant matrices that are already expressed in the "right" (non monomial) basis. This leads us to the problem of studying the generalization of resultant matrices (mainly Bezoutians and companion matrix resultants) to non monomial bases.

Some results on Bezoutians defined with respect to a general basis are found in [139]. A Bézout matrix for Chebyshev polynomials is derived in [8]. Helmke and Fuhrmann's paper on Bezoutians ([69]) addresses the issue of resultant matrices expressed in the control (Horner) basis; moreover, the relationship between the Bézout matrix and the companion matrix resultant in the monomial and Horner basis is further investigated in [29]. Also see [30] for a discussion of polynomial GCD in alternate bases.

A polynomial basis that is widely used in practical applications is the Bernstein basis.

Let Π_n be the space of real polynomials of degree n. Then the basis function of the *Bernstein basis* for Π_n are

$$\phi_i(x) = \binom{n}{i}(1-x)^{n-i}x^i, \qquad i = 0, \ldots n. \qquad (9.3.1)$$

The Bernstein basis is often chosen because it is the most stable basis in the unitary interval $I = [0, 1]$: the condition numbers of the roots in I of an arbitrary polynomial expressed in an arbitrary basis reach their minimum values when the polynomial is expressed in the Bernstein basis (see [52]).

Also, the Bernstein basis is the preferred representation of curves and surfaces in computer-aided geometric design (CAGD) . For example, a planar rational Bézier curve is defined parametrically as $x = x(t)/w(t)$, $y = y(t)/w(t)$, where $x(t)$, $y(t)$ and $w(t)$ are polynomials in the Bernstein basis.

Moreover, it should be pointed out that conversions between the monomial and the Bernstein basis (which can be done either explicitly, or implicitly by a standard parameter substitution) generate very large binomial coefficients, as can be seen from (9.3.1), and are therefore a source of ill conditioning and coefficient growth.

For all these reasons, the study of resultant matrices for Bernstein polynomials is particularly interesting.

The Sylvester matrix for Bernstein polynomials is defined and studied in [133]. A companion matrix and an associated resultant matrix are derived by Winkler in [132]. An analogue of the Bézout matrix for bivariate polynomials expressed in the Bernstein basis, called *Bernstein-Bezoutian matrix*, is considered in [17], whereas the triangular factorization of Bernstein-Bezoutian matrices via a fast fraction-free method is described in [18]. See also [135] and [2] for examples of computation of approximate polynomial GCDs in the Bernstein basis.

9.4. Fast QR decomposition

A promising research direction for fast and stable approximate GCD algorithms employs a new method, devised by Delvaux, Gemignani and Van Barel ([44] and [45]), for the fast QR factorization of a Cauchy-like matrix; recall that resultant matrices can be transformed into Cauchy-like matrices in a fast and stable way ([60]).

Definition 9.4.1. A matrix $B \in \mathbb{C}^{n \times n}$ is called (n_L, n_U)-*quasiseparable* ([48]) if

$$n_L \geq \max_{1 \leq k \leq n-1} \text{rank } B[k+1 : n, 1 : k],$$
$$n_U \geq \max_{1 \leq k \leq n-1} \text{rank } B[1 : k, k+1 : n],$$

where a Matlab-like notation has been used.

When $n_L = n_U = p$, then B is called p-quasiseparable.

A comprehensive review of quasiseparable matrices and applications is found in [128, 129, 130].

The computation of the QR factorization of a Cauchy-like matrix C is reduced to solving an inverse eigenvalue problem for an associate quasiseparable matrix H. Indeed we have:

Theorem 9.4.2. *Let* $C = \left(\frac{\mathbf{g}_i^T \mathbf{h}_j}{d_i^{(1)} - d_j^{(2)}} \right) \in \mathbb{C}^{n \times n}$, *with* $\mathbf{g}_i, \mathbf{h}_j \in \mathbb{C}^p$, *be a Cauchy-like matrix of displacement rank p with respect to the operator*

$$\nabla C = D_1 C - C D_2,$$

where $D_1 = \text{diag}[d_1^{(1)} \ldots d_n^{(1)}]$ *and* $D_2 = \text{diag}[d_1^{(2)} \ldots d_n^{(2)}]$ *are diagonal matrices with mutually distinct entries located on the unit circle or on the real axis. Then there exists a QR factorization of C, C = QR, such that*

$$Q^H D_1 Q = M = D_2 + S, \tag{9.4.1}$$

where S is p-quasiseparable.

This theorem says that the computation of a QR factorization for C can be reduced to determining a matrix S and a unitary matrix Q such that (9.4.1) holds, given in input some information about the structure and the eigensystem of M.

The authors of [44] propose two methods that solve this inverse problem and exploit the quasiseparable structure of M. The second method, in particular, formulates the recursive process of reconstructing the associated quasiseparable matrix as a sequence of QL iterations with ultimate shifts, and is designed as to display good numerical properties. Therefore it may also yield interesting results with respect to the approximate GCD problem.

9.5. A structured approach to the companion matrix resultant

Polynomial GCD computations may also be carried out via the generalized companion matrix introduced in Sections 2.7 and 6.6.3 . Let us recall that, given polynomials $u(x)$ and $v(x)$ with greatest common divisor $g(x)$, the associated generalized companion matrix $v(F_u^T)$ has the following properties (see Lemma 2.7.2):

(i) it is singular if and only if $g(x)$ is a nonconstant polynomial; in particular, $\dim\ker v(F_u^T) = \deg g(x)$;

(ii) $\ker v(F_u^T) = \{\mathbf{z} \in \mathbb{C}^n : z(x) = p(x)w(x) \text{ for some } p(x)\}$.

As a consequence of (i), a suitable rank-revealing factorization of $v(F_u^T)$ allows to detect the degree of $g(x)$. Moreover, (ii) implies that the cofactor $w(x) = u(x)/g(x)$ is, up to multiplication by a scalar, the polynomial of minimum degree defined by a vector in $\ker v(F_u^T)$. In other words, $w(x)$ is the polynomial associated with the "shortest" vector $\mathbf{w} \in \ker v(F_u^T)$, i.e., the equivalence class, with respect to multiplication by a nonzero scalar, of vectors in $\ker v(F_u^T)$ that have a maximum number of trailing zeros.

Observe that the matrix $v(F_u^T)$ has a Toeplitz-like displacement structure of rank 2 with respect to the operator (2.2.5) . One can then devise a fast method for approximate GCD computation that exploits the GKO algorithm, by adapting to the generalized companion matrix the ideas presented in Chapter 7 for the Sylvester or Bézout matrices. It is interesting to point out that the computation of displacement generators for $v(F_u^T)$ only requires to determine a few rows and columns of the matrix itself. This task can be performed cheaply using Barnett's formula (Theorem 2.7.1) : it essentially amounts to solving a structured triangular linear system and computing the product of a Bézout matrix times a vector.

Some preliminary work in this direction, along with a Matlab implementation and some numerical tests, is presented in [21]. The same paper shows how this approach is well-suited to study an asymmetric version of the approximate GCD problem, where only one of the input polynomials is affected by errors, whereas the other polynomial is assumed to be derived from an exact model.

Appendix A
Distances and norms

A.1. Vector norms

Given a positive integer n, let $\mathbf{v} = [v_1, \ldots, v_n] \in \mathbb{C}^n$ and let $1 \le p \le \infty$. We will denote with $\|\mathbf{v}\|_p$ the usual p-norm of $\mathbf{v}$ (which is defined as $\|\mathbf{v}\|_p = (\sum_{i=1}^{n} |v_i|^p)^{1/p}$ when $p < \infty$). In particular:

$$\|\mathbf{v}\|_1 = \sum_{i=1}^{n} |v_i|,$$

$$\|\mathbf{v}\|_2 = \sqrt{\sum_{i=1}^{n} |v_i|^2},$$

$$\|\mathbf{v}\|_\infty = \max_{i=1,\ldots,n} |v_i|.$$

Recall that all vector norms are topologically equivalent, *i.e.*, if $\| \cdot \|_{p_1}$ and $\| \cdot \|_{p_2}$ are two vector norms, then there exist $\alpha, \beta \in \mathbb{R}$, with $0 < \alpha \le \beta$, such that

$$\alpha \|\mathbf{v}\|_{p_2} \le \|\mathbf{v}\|_{p_1} \le \beta \|\mathbf{v}\|_{p_2}$$

for every $\mathbf{v} \in \mathbb{C}^n$. In particular we have:

$$\|\mathbf{v}\|_\infty \le \|\mathbf{v}\|_2 \le \sqrt{n}\|\mathbf{v}\|_\infty,$$
$$\|\mathbf{v}\|_2 \le \|\mathbf{v}\|_1 \le \sqrt{n}\|\mathbf{v}\|_2,$$
$$\|\mathbf{v}\|_\infty \le \|\mathbf{v}\|_1 \le n\|\mathbf{v}\|_\infty.$$

Also recall the *Hölder inequality*:

$$|\mathbf{v}^*\mathbf{w}| \le \|\mathbf{v}\|_p \|\mathbf{w}\|_q, \text{ with } \quad \frac{1}{p} + \frac{1}{q} = 1, \quad \mathbf{v}, \mathbf{w} \in \mathbb{C}^n,$$

a particular case of which is the well-known *Cauchy-Schwarz inequality*:

$$|\mathbf{v}^*\mathbf{w}| \le \|\mathbf{v}\|_2 \|\mathbf{w}\|_2, \quad \mathbf{v}, \mathbf{w} \in \mathbb{C}^n.$$

If $\|\cdot\|$ is a norm on $\mathbb{C}^n$, then $\|\cdot\|_*$ denotes the dual norm on $\mathbb{C}^n$, *i.e.* the norm defined by

$$\|\mathbf{v}\|_* := \max_{\mathbf{w}\neq 0} \frac{|\mathbf{w}^T\mathbf{v}|}{\|\mathbf{w}\|} = \max_{\|\mathbf{w}\|=1} |\mathbf{w}^T\mathbf{v}|.$$

A.2. Matrix norms

Given a vector norm $\|\cdot\|_p$, the associated *induced matrix norm* is defined as

$$\|A\|_p = \max_{\|\mathbf{v}\|_p=1} \|A\mathbf{v}\|_p, \qquad A \in \mathbb{C}^{n\times n}.$$

In particular:

$$\|A\|_1 = \max_{j=1,\dots,n} \sum_{i=1}^n |A(i,j)|,$$

$$\|A\|_2 = \sqrt{\rho(A^*A)},$$

$$\|A\|_\infty = \max_{i=1,\dots,n} \sum_{j=1}^n |A(i,j)|,$$

where $\rho(M)$ denotes the spectral radius of the matrix M.

Another frequently used matrix norm is the *Frobenius norm*:

$$\|A\|_F = \sqrt{\sum_{i,j=1}^n |A(i,j)|^2} = \sqrt{\mathrm{tr}(A^*A)},$$

where $\mathrm{tr}(M)$ denotes the trace of the matrix M.

We also mention the *max norm*:

$$\|A\|_{\max} = \max_{i,j} |A(i,j)|.$$

The following useful inequalities hold for every $A \in \mathbb{C}^{n\times n}$:

$$\frac{1}{\sqrt{n}}\|A\|_\infty \leq \|A\|_2 \leq \sqrt{n}\|A\|_\infty,$$

$$\frac{1}{\sqrt{n}}\|A\|_1 \leq \|A\|_2 \leq \sqrt{n}\|A\|_1,$$

$$\max_{i,j}|A(i,j)| \leq \|A\|_2 \leq n\max_{i,j}|A(i,j)|,$$

$$\|A\|_2 \leq \sqrt{\|A\|_1 \cdot \|A\|_\infty},$$

$$\|A\|_2 \leq \|A\|_F \leq \sqrt{n}\|A\|_2.$$

A.3. Polynomial metrics

Let $\mathcal{P}_n$ be the set of polynomials with complex coefficients and degree at most n. A polynomial $p(z) \in \mathcal{P}_n$ will be denoted through its coefficients as

$$p(z) = \sum_{j=0}^{n} p_j z^j.$$

We seek to assign a metric d on $\mathcal{P}_n$. Of course there are many ways to do this; the choice usually depends on which features of the analyzed problem should be highlighted. For most purposes related to the GCD problem, a convenient candidate for d belongs to one of the following two classes.

1. Let $1 \leq s \leq \infty$. For $p(z), q(z) \in \mathcal{P}_n$, let $\mathbf{p}, \mathbf{q} \in \mathbb{C}^{n+1}$ be the associated vectors and set

$$d(p, q) = \|p - q\|_s := \|\mathbf{p} - \mathbf{q}\|_s.$$

   In other words, this is the metric induced by the usual s-norm on $\mathbb{C}^{n+1}$.

2. Choose a vector $\mathbf{m}$ of positive real numbers $m_j \geq 0$. Using the same notation as above, define

$$d(p, q) = |p - q|_{\mathbf{m}} := \max_j \frac{|p_j - q_j|}{m_j},$$

   so that d is the distance induced by the usual ∞-norm on $\mathbb{C}^{n+1}$ with weights $1/m_0, \ldots, 1/m_n$.

The choice of the weights m_j determines the relative importance of the variation in each coefficient. For example, choosing $m_j = 1$ for all $j = 1, \ldots, n$ means considering the absolute differences between the coefficients. But if a particular polynomial $p(z)$ is being studied, it is a common choice to consider relative differences with respect to the coefficients of $p(z)$, i.e. set $m_j = |p_j|$.

For some purposes it can be advantageous to use nonnegative – rather than strictly positive – weights, with the convention that $1/0 = \infty$ and $0/0 = 0$. This choice allows to preserve a possible sparse structure of $p(z)$, or more generally to study cases where only some of the coefficients are perturbed. More information on polynomial norms and inequalities can be found in [9] and [150].

Appendix B
Special matrices

B.1. Convolution matrices

Polynomial multiplication can be expressed in matrix form via *convolution matrices*. We assume here to work with complex polynomials.

B.1.1. Univariate case

Let $a(x) = \sum_{i=0}^{n} a_i x^i$ and $b(x) = \sum_{i=0}^{m} b_i x^i$ be polynomials with given coefficients; let $\mathbf{a}$ and $\mathbf{b}$ be the associated vectors. Consider the product $c(x) = a(x)b(x)$. Then the associated vector $\mathbf{c}$ is given by

$$\mathbf{c} = \mathcal{C}_{m+1}(a) \cdot \mathbf{b}, \tag{B.1.1}$$

where

$$\mathcal{C}_{m+1}(a) = \begin{pmatrix} a_0 & 0 & \cdots & 0 \\ a_1 & a_0 & \ddots & \vdots \\ \vdots & a_1 & \ddots & 0 \\ \vdots & \vdots & & a_0 \\ a_n & \vdots & & a_1 \\ 0 & a_n & & \vdots \\ \vdots & \ddots & \ddots & \vdots \\ 0 & \cdots & 0 & a_n \end{pmatrix}$$

is called the $(m+1)$-st convolution matrix associated with the polynomial $a(x)$. Observe that $\mathcal{C}_{m+1}(a)$ is an $(n+m+1) \times (m+1)$ Toeplitz matrix.

In other words, multiplication by $a(x)$ can be seen, for any m (degree of the second factor) as a linear transformation

$$\mathcal{C}_{m+1}(a) : \mathbb{C}^{m+1} \longrightarrow \mathbb{C}^{n+m+1}$$

whose associated matrix is the $(m+1)$-st convolution matrix (we use here the same symbol for a linear transformation and its associated matrix).

Since polynomial multiplication is commutative, we may also write

$$\mathbf{c} = \mathcal{C}_{n+1}(b) \cdot \mathbf{a},$$

where $\mathcal{C}_{n+1}(b)$ is the $(n+1)$-th convolution matrix associated with $b(x)$.

This matrix formulation is also useful, from a numerical point of view, to perform polynomial division in a stable way, when a zero remainder is expected. Indeed, if polynomials $a(x)$ and $c(x)$ are given, then the coefficients of $b(x) = c(x)/a(x)$ can be computed by solving the system (B.1.1) in a least squares sense.

B.1.2. Multivariate case

The notion of convolution matrix can be generalized to the multivariate case.

Let us first fix the notation for the multivariate case. Let $u(x_1, \ldots, x_r)$ be an r-variate polynomial. The r-degree of u is $n = [n_1, \ldots, n_r]$, where, n_i is the degree of u with respect to the variable x_i, for $i = 1, \ldots, r$. For example, the bivariate polynomial

$$u(x_1, x_2) = x_1^4 x_2 - 5x_1^2 x_2^3 - 1$$

has 2-degree $[4, 3]$.

If two polynomials have r-degrees $n = [n_1, \ldots, n_r]$ and $m = [m_1, \ldots, m_r]$, we say that $m \le n$ if $m_i \le n_i$ for $i = 1, \ldots, r$. The sum of r-degrees is defined as $n + m = [n_1 + m_1, \ldots, n_r + m_r]$.

The vector space $\mathbb{P}_n$ of all r-variate polynomials of degree less or equal to n has dimension $v(n) = \prod_{i=0}^{r}(n_i + 1)$. The monomial basis for $\mathbb{P}_n$ is

$$\mathcal{B} = \left\{ x_1^{k_1} x_2^{k_2} \ldots x_r^{k_r} \middle| [k_1, k_2 \ldots, k_r] \le n \right\}.$$

Choosing an order on the elements of $\mathcal{B}$ (for example, the lexicographic order) allows to associate a vector $\mathbf{u}$ of coefficients with any polynomial $u \in \mathbb{P}_n$.

Now, assume that $a(x_1, \ldots, x_r)$ and $b(x_1, \ldots, x_r)$ are r-variate polynomials whose coefficients with respect to the power basis are given. Let $[n_1, \ldots, n_r]$ and $[m_1, \ldots, m_r]$ be the r-degrees of $a(x_1, \ldots, x_r)$ and $b(x_1, \ldots, x_r)$, respectively, and let $\mathbf{a}$ and $\mathbf{b}$ be the associated vectors.

As in the univariate case, multiplication by a is a linear transformation

$$\mathcal{C}_m(a) : \mathbb{P}_m \longrightarrow \mathbb{P}_{n+m}$$

whose associated matrix $C_m(a)$ has size $v(n+m) \times v(m)$. If $\mathcal{B} = (\gamma_1, \ldots, \gamma_{v(m)})$ is the ordered monomial basis of $\mathbb{P}_m$ and we set $b(j) = b \cdot \gamma_j$ for $j = 1, \ldots, v(m)$, then the columns of the matrix $C_m(a)$ are given by the vectors associated with the $b(j)$'s:

$$C_m(a) = [\mathbf{b}^{(1)} \quad \mathbf{b}^{(2)} \quad \ldots \quad \mathbf{b}^{(v(m))}].$$

As in the univariate case, of course, the roles of a and b are interchangeable.

B.2. The Fourier matrix and polynomial computations

For a fixed integer n, let $\omega = e^{\frac{2\pi i}{n}}$ and define the *Fourier matrix* of order n as the $n \times n$ complex matrix $\mathcal{F}$ such that

$$\mathcal{F}^*(i, j) = \frac{1}{\sqrt{n}} \omega^{(i-1)(j-1)}$$

or, more explicitly:

$$\mathcal{F}^* = \frac{1}{\sqrt{n}} \begin{pmatrix} 1 & 1 & 1 & \cdots & 1 \\ 1 & \omega & \omega^2 & \cdots & \omega^{n-1} \\ 1 & \omega^2 & \omega^4 & \cdots & \omega^{n-2} \\ \vdots & \vdots & \vdots & & \vdots \\ 1 & \omega^{n-1} & \omega^{n-2} & \cdots & \omega \end{pmatrix}.$$

Observe that there are only n distinct entries in $\mathcal{F}$.

Theorem B.2.1. *The Fourier matrix is unitary.*

Proof. The desired relations

$$\mathcal{F}\mathcal{F}^* = \mathcal{F}^*\mathcal{F} = I$$

are a consequence of the geometric series identity

$$\sum_{h=0}^{n-1} \omega^{h(j-k)} = \frac{1 - \omega^{n(j-k)}}{1 - \omega^{j-k}} = \begin{cases} n & \text{if } j = k \\ 0 & \text{if } j \neq k \end{cases}. \qquad \square$$

Let $\mathbf{u} \in \mathbb{C}^n$. Then

$$\hat{\mathbf{u}} = \mathcal{F}\mathbf{u}$$

is the *discrete Fourier transform* (DFT) of $\mathbf{u}$. Its inverse is given by

$$\mathbf{u} = \mathcal{F}^*\hat{\mathbf{u}}.$$

The DFT assumes a special meaning when $\mathbf{u}$ is the coefficient vector associated with a polynomial $u(x)$. More precisely, we have

$$\sqrt{n}\mathcal{F}^* \begin{bmatrix} u_0 \\ u_1 \\ \vdots \\ u_{n-1} \end{bmatrix} = \begin{bmatrix} u(1) \\ u(\omega) \\ \vdots \\ u(\omega^{n-1}) \end{bmatrix},$$

that is, the IDFT (inverse DFT) applied to $\mathbf{u}$ yields the values taken by $u(x)$ at the n-th roots of unity. Conversely, we have

$$\begin{bmatrix} u_0 \\ u_1 \\ \vdots \\ u_{n-1} \end{bmatrix} = \frac{1}{\sqrt{n}} \begin{bmatrix} u(1) \\ u(\omega) \\ \vdots \\ u(\omega^{n-1}) \end{bmatrix},$$

In other words, the DFT and its inverse can be used to evaluate or interpolate a polynomial. They can be computed in a fast (*i.e.* with $\mathcal{O}(n \log n)$ ops) and stable way via FFT (fast Fourier transform) and can be used to accelerate polynomial multiplication and division (see [19]).

Problem B.2.2. (Multiplication of two polynomials, or convolution of two vectors.) Given the input values $u_0, \ldots, u_n, v_0, \ldots, v_m$, compute

$$w_i = \sum_{j=0}^{i} u_j v_{i-j}, \qquad i = 0, 1, \ldots, m+n,$$

where we assume $u_j = 0$ for $j > n$ and $v_k = 0$ for $k > m$. Observe that the w_i's are the coefficients of the polynomial $w(x) = u(x)v(x)$, with $u(x) = \sum_{i=0}^{n} u_i x^i$ and $v(x) = \sum_{i=0}^{m} v_i x^i$. Equivalently, $\mathbf{w}$ is the convolution of $\mathbf{u}$ and $\mathbf{v}$.

The problem can be solved using the FFT, as follows:

- choose $K = 2^k$, for an integer k, such that $m + n \leq K < 2(m+n)$;
- evaluate $u(x)$ and $v(x)$ on the K-th roots of unity $1, \omega, \ldots, \omega^{K-1}$;
- compute $w(\omega^h) = u(\omega^h) \cdot v(\omega^h)$ for $h = 0, 1, \ldots, K-1$;
- recover the coefficients of $w(x)$ from its values on the K-th roots of unity.

Problem B.2.3. (Exact polynomial division.)   Given the input values $u_0, \ldots, u_n, v_0, \ldots, v_m$, where the polynomial $v(x) = \sum_{i=0}^{m} v_i x^i$ divides exactly $u(x) = \sum_{i=0}^{n} u_i x^i$, compute the coefficients $w_0, \ldots, w_{n-m}$ of $w(x) = u(x)/v(x)$.

Again, this problem can be solved using an evaluation/interpolation technique:

- let $K = n - m + 1$; evaluate $u(x)$ and $v(x)$ on the K-th roots of unity;
- compute $w(\omega^h) = u(\omega^h)/v(\omega^h)$ for $h = 0, 1, \ldots, K - 1$;
- recover the coefficients of $w(x)$ from its values on the K-th roots of unity.

Observe that some details in the above method need to be taken care of, such as the case when $v(\omega^h) = 0$ for some value of h.

B.3. Circulant matrices

A *circulant matrix* of order n is a square matrix of the form

$$
C = \begin{pmatrix}
c_1 & c_2 & \cdots & c_{n-1} & c_n \\
c_n & c_1 & \cdots & c_{n-2} & c_{n-1} \\
\vdots & & \ddots & & \vdots \\
c_3 & c_4 & \cdots & c_1 & c_2 \\
c_2 & c_3 & \cdots & c_n & c_1
\end{pmatrix}.
$$

In other words, a circulant matrix is a special Toeplitz matrix such that its diagonals "wrap around". The whole matrix is completely determined by its first row (or column) and therefore depends only on n parameters. Observe that the circulant matrices form a linear subspace of the space of $n \times n$ real or complex matrices.

We state here some useful facts about circulant matrices; see [43] for a detailed presentation.

Let

$$
\Pi = \begin{pmatrix}
0 & 1 & 0 & \cdots & 0 \\
\vdots & \ddots & \ddots & & \vdots \\
\vdots & & & \ddots & 1 & 0 \\
0 & \cdots & & \cdots & 0 & 1 \\
1 & 0 & \cdots & \cdots & 0
\end{pmatrix},
$$

i.e., $\Pi = Z_1^T$ with the notation of (2.2.6). Circulant matrices can be characterized as the matrices that commute with Π:

Theorem B.3.1. *Let A be an $n \times n$ matrix. Then A is circulant if and only if $A\Pi = \Pi A$.*

It follows from the above result that A is circulant if and only if A^* is circulant.

In view of the structure of the matrices $\Pi, \Pi^2, \ldots, \Pi^{n-1}$, it is clear that the matrix C defined above can be written as

$$C = c_1 \mathbf{I} + c_2 \Pi + c_3 \Pi^2 + \cdots + c_n \Pi^{n-1}.$$

Therefore, an $n \times n$ matrix is circulant if and only if it can be written as $p(\Pi)$ for some polynomial $p(z)$ of degree $n - 1$.

Associate with the n-tuple $\gamma = (c_1, c_2 \ldots, c_n)$ the polynomial

$$p_\gamma(z) = c_1 + c_2 z + \cdots + c_n z^{n-1}.$$

The polynomial $p_\gamma(z)$ is called the *representer* of the associated circulant matrix. The association $\gamma \leftrightarrow p_\gamma(z)$ is clearly linear.

Since polynomials in the same matrix commute, it follows that all circulants of the same order commute (in particular, observe that C and C^* commute and therefore all circulant matrices are normal). We conclude that the set of circulant matrices of order n is a commutative algebra with generator matrix Π.

Circulant matrices have the very important property that they are diagonalized by the Fourier matrix.

We start by diagonalizing the matrix Π. Let $\omega = e^{\frac{2\pi i}{n}}$ and define the diagonal matrix

$$\Omega = \operatorname{diag}(1, \omega, \omega^2, \ldots, \omega^{n-1}).$$

Then, by a straightforward application of the definitions, we have:

$$\Pi = \mathcal{F}^* \Omega \mathcal{F}.$$

It follows that

Theorem B.3.2. *Let C be a circulant matrix. Then*

$$C = \mathcal{F}^* \Lambda \mathcal{F},$$

where

$$\Lambda = \operatorname{diag}(p_\gamma(1), p_\gamma(\omega), \ldots, p_\gamma(\omega^{n-1})).$$

Proof. We have:

$$
\begin{aligned}
C &= p_\gamma(\Pi) \\
&= p_\gamma(\mathcal{F}^* \Omega \mathcal{F}) \\
&= \mathcal{F}^* p_\gamma(\Pi) \mathcal{F} \\
&= \mathcal{F}^* \cdot \operatorname{diag}(p_\gamma(1), p_\gamma(\omega), \ldots, p_\gamma(\omega^{n-1})). \qquad \square
\end{aligned}
$$

Conversely, we have:

Theorem B.3.3. *Let*

$$\Lambda = \text{diag}(p_\gamma(1), p_\gamma(\omega), \ldots, p_\gamma(\omega^{n-1})).$$

Then

$$C = \mathcal{F}^* \Lambda \mathcal{F},$$

is a circulant matrix.

A generalization of circulant matrices is given by f-*circulant* matrices. Let $f \in \mathbb{C}$; then an f-circulant matrix has the form

$$C_f = \begin{pmatrix} c_1 & fc_2 & \cdots & fc_{n-1} & fc_n \\ c_n & c_1 & \cdots & fc_{n-2} & fc_{n-1} \\ \vdots & & \ddots & & \vdots \\ c_3 & c_4 & \cdots & c_1 & fc_2 \\ c_2 & c_3 & \cdots & c_n & c_1 \end{pmatrix},$$

i.e., it is a circulant matrix whose strictly upper triangular part has been multiplied by f. (-1)-circulant matrices are called *anticirculant* or *skew-circulant*.

The properties of circulant matrices seen so far can be generalized to f-circulant matrices (see [43], [19], [37]). In particular, we have:

Theorem B.3.4. *Let* $D_f = \text{diag}(1, g, g^2, \ldots, g^{n-1})$, *where* $g^n = f$. *Let* $\mathbf{c}^T$ *denote the first row of an* $n \times n$ f-*circulant matrix* C_f. *Then*

$$\mathcal{F} D_f C_f D_f^{-1} \mathcal{F}^* = D,$$

with $D = \text{diag}(d_0, \ldots, d_{n-1})$, *where the* d_i's *are the entries of* $\mathbf{d} = \sqrt{n} \mathcal{F} D_f \mathbf{c}$.

References

[1] A. V. AHO, J.E. HOPCROFT and J. D. ULLMANN, "The Design and Analysis of Computer Algorithms", Addison-Wesley Series in Computer Science and Information Processing, Addison-Wesley Publishing Co., Reading, Massachusetts, 1975.

[2] J. D. ALLAN and J. R. WINKLER, *Structure preserving methods for the computation of approximate GCDs of Bernstein polynomials*, In: P. Chenin, T. Lyche and l. L. Schumacher, editors, "Curve and Surface Design: Avignon 2006", 11-20, Nashboro Press, Tennessee, 2007.

[3] A. AMIRASLANI, R. M. CORLESS, L. GONZALEZ-VEGA and A. SHAKOORI, "Polynomial Algebra by Values", Technical Report TR-04-01, Ontario Research Centre for Computer Algebra (January 2004).

[4] A. ARICÒ and G. RODRIGUEZ, *A fast solver for linear systems with displacement structure*, Numer. Alg. **55(4)**(2010), 529–556.

[5] S. BARNETT, *Greatest common divisor of several polynomials*, Proc. Cambridge Philos. Soc. **70** (1971), 263–268.

[6] S. BARNETT, *A Note on the Bezoutian Matrix*, SIAM J. Appl. Math. **22**(1972), 84–86.

[7] S. BARNETT, "Polynomials and Linear Control Systems", Monographs and textbooks in pure and applied mathematics, Marcel Dekker Inc., New York, 1983.

[8] S. BARNETT, "A Bezoutian Matrix for Chebyshev polynomials", IMA Conference on Application of Matrix Theory, The University of Bradford, England, 1988.

[9] B. BEAUZAMY, E. BOMBIERI, P. ENFLO and H. L. MONTGOMERY, *Products of polynomials in many variables*, J. Number Theory **36(2)** (1990), 219–245.

[10] B. BECKERMANN and G. LABAHN, *When are two numerical polynomials relatively prime?*, J. Symbolic Comput. **26(6)** (1998), 677-689.

[11] B. BECKERMANN and G. LABAHN, *A fast and numerically stable Euclidean-like algorithm for detecting relatively prime numerical polynomials*, J. Symb. Comp. **26(6)** (1998), 691–714.

[12] D. A. BINI and P. BOITO, *Structured matrix based methods for polynomial ϵ-GCD: analysis and comparisons*, Proc.of the 2007 International Symposium on Symbolic and Algebraic Computation (Waterloo, ON), ACM Press (2007), 9–16.

[13] D. A. BINI and P. BOITO, *A fast algorithm for approximate polynomial gcd based on structured matrix computations*, In: "Numerical Methods for Structured Matrices and Applications: Georg Heinig memorial volume, Operator Theory: Advances and Applications", vol. 199, Birkhäuser, 155–173, 2010.

[14] D. A. BINI and G. FIORENTINO, *Design, analysis and implementation of a multiprecision polynomial rootfinder*, Numerical Algorithms **23(2-3)** (2000), 127–173.

[15] D. A. BINI and L. GEMIGNANI, *Fast parallel computation of the polynomial remainder sequence via Bézout and Hankel matrices*, SIAM J. Comput. **24(1)** (1995), 63–77.

[16] D. A. BINI and L. GEMIGNANI, *Fast fraction-free triangularization of Bezoutians with applications to sub-resultant chain computation*, Linear Algebra Appl. **284(1-3)** (1998), 19–39.

[17] D. A. BINI and L. GEMIGNANI, *Bernstein-Bezoutian matrices*, Theoret. Comput. Sci. **315(2-3)** (2004), 319-333.

[18] D. A. BINI, L. GEMIGNANI and J. WINKLER, *Structured matrix methods for CAGD: An application to computing the resultant of polynomials in Bernstein basis*, Numer. Linear Algebra Appl. **12(8)** (2005), 685–698.

[19] D. A. BINI and V. Y. PAN, "Polynomial and Matrix Computations", Vol. I: Fundamental Algorithms, Progress in Theoretical Computer Science, Birkhäuser, Boston, 1994.

[20] C. H. BISCHOF, *Incremental condition estimation*, SIAM J. Matrix Anal. Appl. **11(4)** (1990), 312-322.

[21] P. BOITO and O. RUATTA, *Generalized companion matrix for approximate GCD*, arXiv:1102.1809v1 [cs.SC] (2011).

[22] A. BÖTTCHER and S. M. GRUDSKY, *Structured condition numbers of large Toeplitz matrices are rarely better than usual condition numbers*, Numer. Linear Algebra Appl. **12(2-3)** (2005), 95–102.

[23] R. P. BRENT, *Stability of fast algorithms for structured linear systems*, In: "Fast Reliable Algorithms for Matrices with Structure", 103–116, SIAM, Philadelphia, PA, 1999.

[24] R. B. BRENT, F. G. GUSTAVSON and D. Y. Y. YUN, *Fast so-lution of Toeplitz systems of equations and computation of Padé approximations*, J. Algorithms **1(3)** (1980), 259–295.

[25] W. S. BROWN, *On Euclid's algorithm and the computation of polynomial greatest common divisors*, J. Assoc. Comput. Mach. **18** (1971), 478–504.

[26] W. S. BROWN and J. F. TRAUB, *On Euclid's Algorithm and the Theory of Subresultants*, J. Assoc. Comput. Mach. **18** (1971), 505–514.

[27] J. R. BUNCH, *Stability of Methods for Solving Toeplitz Systems of Equations*, SIAM J. Sci. Stat. Comput. **6(2)** (1985), 349–364.

[28] S. CABAY and R. MELESHKO, *A weakly stable algorithm for Padé approximants and the inversion of Hankel matrices*, SIAM J. Matrix Anal. Appl. **14(3)** (1993), 735–765.

[29] J.-P. CARDINAL, "On Two Iterative Methods for Approximating the Roots of a Polynomial", Lectures in Applied Mathematics, Vol. 32, Amer. Math. Soc., Providence, RI, 1996.

[30] H. CHENG and G. LABAHN, *On Computing Polynomial GCD in Alternate Bases*, Proceedings of the 2006 International Symposium on Symbolic and Algebraic Computation (Genova, Italy), 47–54, ACM Press (2006).

[31] G. CHÈZE, A. GALLIGO, B. MOURRAIN and J.-C. YAKOUB-SOHN, *Computing Nearest Gcd with Certification*, Proceedings of Symbolic Numeric Computation 2009, 29-34, ACM Press, New York, 2009.

[32] G. CHÈZE, A. GALLIGO, B. MOURRAIN and J.-C. YAKOUB-SOHN, *A Subdivision Method for Computing Nearest Gcd with Certification*, preprint, 2011.

[33] P. CHIN, R. M. CORLESS and G. F. CORLISS, *Optimization Strategies for the Approximate GCD Problem*, Proceedings of the 1998 International Symposium in Symbolic and Algebraic Computation (Rostock), 228-235, ACM Press, New York (1998).

[34] D. CHRISTOU, N. KARCANIAS and M. MITROULI, *The ERES method for computing the approximate GCD of several polynomials*, Appl. Numer. Math. **60** (2010), 94–114.

[35] M. T. CHU, R. E. FUNDERLIC and R. J. PLEMMONS, *Structured low rank approximation*, Linear Algebra Appl. **366** (2003), 157–172.

[36] J. CHUN, T. KAILATH and H. LEV-ARI, *Fast parallel algorithms for QR and triangular factorization*, SIAM J. Sci. Stat. Comp. **8(6)** (1987), 899–913.

[37] R. E. CLINE, R. J. PLEMMONS and G. WORM, *Generalized inverses of certain Toeplitz matrices*, Linear Algebra Appl. **8** (1974), 25–33.

[38] G. E. COLLINS, *Subresultants and reduced polynomial remainder sequences*, J. Assoc. Comput. Mach. **14** (1967), 128–142.

[39] R. M. CORLESS, *Editor's corner: Cofactor iteration*, SIGSAM Bulletin **30** (1996), 34–38.

[40] R. M. CORLESS, P. M. GIANNI, B. M. TRAGER and S. M. WATT,*The Singular Value Decomposition for Approximate Polynomial Systems*, Proceeedings of the 1995 International Symposium on Symbolic and Algebraic Computation (Montreal, Canada), 195–207, ACM Press (1995).

[41] R. M. CORLESS, STEPHEN M. WATT and LIHONG ZHI, *QR Factoring to Compute the GCD of Univariate Approximate Polynomials*, IEEE Trans. Signal Processing **52(12)** (2004), 3394–3402.

[42] G. CYBENKO, *The Numerical Stability of the Levinson-Durbin Algorithm for Toeplitz Systems of Equations*, SIAM J. Sci. Stat. Comput. **1(3)** (1980), 303–310.

[43] P. J. DAVIS, "Circulant Matrices", Wiley Interscience, Pure and Applied Mathematics, John Wiley & Sons, New York-Chichester-Brisbane, 1979.

[44] S. DELVAUX, L. GEMIGNANI and M. VAN BAREL, *Fast QR Factorization of Cauchy-like Matrices*, Linear Algebra Appl. **428** (2008), 697–711.

[45] S. DELVAUX, L. GEMIGNANI and M. VAN BAREL, *QR-factorization of displacement structured matrices using a rank structured matrix approach*, In: "Numerical Methods for Structured Matrices and Applications: Georg Heinig memorial volume, Operator Theory: Advances and Applications", Vol. 199, Birkhäuser, 229–254, 2010.

[46] G. M. DIAZ-TOCA and L. GONZALEZ-VEGA, *Barnett's theorems about the greatest common divisor of several univariate polynomials through Bézout-like matrices*, J. Symbolic Comput. **34(1)** (2002), 59–81.

[47] G. M. DIAZ-TOCA and L. GONZALEZ-VEGA, *Computing greatest common divisors and squarefree decompositions through matrix methods: The parametric and approximate cases*, Linear Algebra Appl. **412(2-3)** (2006), 222–246.

[48] Y. EIDELMAN and I. GOHBERG, *On a new class of structured matrices*, Integral Equations Operator Theory **34(3)** (1999), 293–324.

[49] I. Z. EMIRIS, A. GALLIGO and H. LOMBARDI, *Certified approximate univariate GCDs*, J. Pure Appl. Algebra **117/118** (1997), 229–251.

[50] I. Z. EMIRIS, A. GALLIGO, H. LOMBARDI, *Numerical Univariate Polynomial GCD*, in: J. Renegar, M. Shub and S. Smale, eds., The Mathematics of Numerical Analysis, Lectures in Applied Mathematics **32** (AMS, Providence, RI, 1996), 323-343.

[51] J. ERIKSSON, P. A. WEDIN, M. E. GULLIKSSON and I. SÖDERKVIST, *Regularization Methods for Uniformly Rank-Deficient Nonlinear Least-Squares Problems*, J. Optim. Theory Appl. **127(1)** (2005), 1–26.

[52] R. T. FAROUKI and T. N. T. GOODMAN, *On the optimal stability of the Bernstein basis*, Math. Comput. **65(216)** (1993), 1553–1566.

[53] R. FLETCHER, *Generalized inverse methods for the best least squares solution of systems of non-linear equations*, Comput. J. **10** (1967/68), 392 399.

[54] R. FLETCHER, "Practical methods of Optimization Vol. I: Unconstrained Optimization", Wiley-Interscience, John Wiley & Sons Ltd., Chichester, 1981.

[55] P. A. FUHRMANN, "A polynomial approach to linear algebra", Universitext, Springer Verlag, New York, 1996.

[56] Z. GALIL and V. Y. PAN, *Improved processor bounds for combinatorial problems in RNC*, Combinatorica **8(2)** (1988), 189–200.

[57] S. GAO, E. KALTOFEN, J. MAY, Z. YANG and L. ZHI, *Approximate Factorization of Multivariate Polynomials via Differential Equations*, Proceedings of the 2004 International Symposium on Symbolic and Algebraic Computation, 167–174, ACM Press, New York, 2004.

[58] K. O. GEDDES, S. R. CZAPOR and G. LABAHN, "Algorithms for Computer Algebra", Kluwer Academic Publishers, Boston, MA, 1992.

[59] L. GEMIGNANI, "GCD of Polynomials and Bézout Matrices", Proceedings of the 1997 International Symposium on Symbolic and Algebraic Computation (Kihei, Maui, HI), 271-277, ACM Press, New York, 1997.

[60] I. GOHBERG, T. KAILATH and V. OLSHEVSKY, *Fast Gaussian elimination with partial pivoting for matrices with displacement structure*, Math. Comp. **64(12)** (1995), 1557–1576.

[61] I. GOHBERG and A. SEMENCUL, *The inversion of finite Toeplitz matrices and their continual analogues*, Mat. Issled. **7(2)** (1972), 201–223.

[62] G. H. GOLUB and C. F. VAN LOAN, "Matrix Computations", 3rd edition, The Johns Hopkins Studies in the Mathematical Sciences, The Johns Hopkins University Press, Baltimore, MD, 1996.

[63] S. GRAILLAT and P. LANGLOIS, "Testing Polynomial Primality with Pseudozeros", Proc. Fifth Conference on Real Numbers and Computers, Lyon, France, 2003.

[64] G. HEINIG and K. ROST, *Algebraic Methods for Toeplitz-like Matrices and Operators*, Birkhäuser, Basel, 1984.

[65] G. HEINIG, *Inversion of generalized Cauchy matrices and other classes of structured matrices*, Linear Algebra in Signal Processing, IMA volumes in Mathematics and its Applications **69** (1994), 95–114.

[66] G. HEINIG, *Matrix Representations of Bezoutians*, Linear Algebra Appl. **223/224** (1995), 337–354.

[67] G. HEINIG, P. JANKOWSKY and K. ROST, *Fast inversion of Toeplitz-plus-Hankel matrices*, Numer. Math. **52(6)** (1988), 665–682.

[68] G. HEINIG and K. ROST, "Algebraic Methods for Toeplitz-like Matrices and Operators", Mathematical Research, Vol. 19, Akademie Verlag, Berlin and Operator Theory: Advances and Applications **13**, Birkhäuser, Basel, 1984.

[69] U. HELMKE and P. A. FUHRMANN, *Bezoutians*, Linear Algebra Appl. **124** (1989), 1039–1097.

[70] N. J. HIGHAM, "Accuracy and Stability of Numerical Algorithms", 2nd edition, SIAM, Philadelphia, PA, 2002.

[71] J. E. HOPCROFT and R. M. KARP, *An $n^{5/2}$ algorithm for maximum matching in bipartite graphs*, SIAM J. Computing **2** (1973), 225–231.

[72] D. G. HOUGH, "Explaining and Ameliorating the Ill Condition of the Zeros of Polynomials", Ph.D. Thesis, Comp. Science Div., Univ. of California, Berkeley, 1977.

[73] V. HRIBERNIG, "Sensitivity of Algebraic Algorithms", Ph.D. Thesis at the Technical University of Vienna, 1994.

[74] V. HRIBERNIG and H. J. STETTER, *Detection and validation of clusters of polynomial zeros*, J. Symb. Comp. **24(6)** (1997), 667–681.

[75] C.-P. JEANNEROD and G. LABAHN, "SNAP User's Guide", University of Waterloo Technical Report no. CS-2002-22, 2002.

[76] T. KAILATH and J. CHUN, *Generalized displacement structure for block-Toeplitz, Toeplitz-block, and Toeplitz-derived matrices*, SIAM J. Matrix Anal. Appl. **15(1)** (1994), 114–128.

[77] T. KAILATH, S. Y. KUNG and M. MORF, *Displacement ranks of a matrix*, Bull. Amer. Math. Soc. **1(5)** (1979), 769–773.

[78] T. KAILATH and A. H. SAYED, "Fast Algorithms for Generalized Displacement Structures", Recent Advances in Mathematical Theory of Systems, Control, Networks and Signal Processing II, Proc. MTNS-91 (H. Kimura, S. Kodama eds.), Mita Press, Japan, 1992, pp. 27–32.

[79] T. KAILATH and A. H. SAYED, *Displacement Structure: Theory and Applications*, SIAM Review **37(3)** (1995), 297–386.

[80] E. KALTOFEN, Z. YANG and L. ZHI, "Structured Low Rank Approximation of a Sylvester Matrix", In: Dongming Wang and Lihong Zhi (eds.), Proc. International Workshop on Symbolic-Numeric Computation, 2005.

[81] E. KALTOFEN, Z. YANG and L. ZHI, "Approximate Greatest Common Divisors of Several Polynomials with Linearly Constrained Coefficients and Simgular Polynomials", Proceedings of the 2006 International Symposium on Symbolic and Algebraic Computations (Genova, Italy), ACM Press, 2006.

[82] N. KARKANIAS and M. MITROULI, *A Matrix Pencil Based Numerical Method for the Computation of the GCD of Polynomials*, IEEE Trans. Automatic Control **39(5)** (1994), 977–981.

[83] N. KARKANIAS, S. FATOUROS, M. MITROULI and G. H. HA-LIKIAS, *Approximate Greatest Common Divisor of Many Polynomials, Generalised Resultants, and Strength of Approximation*, Comp. Math. Appl. **51(12)** (2006), 1817–1830.

[84] N. K. KARMARKAR and Y. N. LAKSHMAN, *Approximate polynomial greatest common divisors and nearest singular polynomials*, Proc. Int. Symp. Symbolic Algebraic Comput., Zürich, Switzerland, 1996, pp. 35–42.

[85] N. K. KARMARKAR and Y. N. LAKSHMAN, *On Approximate GCDs of Univariate Polynomials*, J. Symbolic Comp. 26(6) (1998), 653–666.

[86] P. KIRRINNIS, *Partial Fraction Decomposition in* $\mathbb{C}[z]$ *and Simultaneous Newton Iteration for Factorization in* $\mathbb{C}[z]$, J. Complexity **14(3)** (1998), 378–444.

[87] M. GU, *Stable and Efficient Algorithms for Structured Systems of Linear Equations*, SIAM J. Matrix Anal. Appl. **19(2)** (1998), 279–306.

[88] M. A. LAIDACKER, *Another theorem relating Sylvester's matrix and the greatest common divisor*, Math. Mag. **42** (1969), 126–128.

[89] P. LANCASTER and M. TISMENETSKY, "The Theory of Matrices", 2nd edition, Computer Science and Applied Mathematics, Academic Press Inc., Orlando, FL, 1985.

[90] B. LI, J. NIE and L. ZHI, *Approximate GCDs of polynomials and sparse SOS relaxation*, Theor. Comput. Sci. **409** (2008), 200–210.

[91] B. LI, Z. YANG and L. ZHI, *Fast Low Rank Approximation of a Sylvester Matrix by Structure Total Least Norm*, Journal of Japan Society for Symbolic and Algebraic Computation **11** (2005), 165–174.

[92] E. LINZER, "Can Symmetric Toeplitz Solvers be Strongly Stable?", Proc. ISTCS '92, (D. Dolev, Z. Galil, M. Rodeh, eds.), Lecture Notes in Computer Science, Vol. 601, 137–146, Springer, Berlin, 1992.

[93] E. LINZER and M. VETTERLI, *Iterative Toeplitz Solvers with Local Quadratic Convergence*, Computing **49(4)** (1993), 339–347.

[94] R. LOOS, "Generalized Polynomial Remainder Sequences", Computer Algebra: Symbolic and Algebraic Computation (Wien) (B. Buchberger, G. E. Collins, R. Loos eds.), Springer-Verlag, Wien, 1982.

[95] I. MARKOVSKY and S. VAN HUFFEL, *An algorithm for approximate common divisor computation*, In: Proc. of the 17th Symposium on Mathematical Theory of Networks and Systems, 2006, 274–279.

[96] I. MARKOVSKY, S. VAN HUFFEL and A. KUKUSH, *On the computation of the multivariate structured total least squares estimator*, Numer. Linear Algebra Appl. **11(5-6)** (2004), 591–608.

[97] I. MARKOVSKY, S. VAN HUFFEL and R. PINTELON, *Block Toeplitz-Hankel structured total least squares*, SIAM J. Matrix Anal. Appl. **26(4)** (2005), 1083–1099.

[98] R. G. MOSIER, *Root neighborhoods of a polynomial*, Math. Comp. **47(175)** (1986), 265–273.

[99] J. NIE, J. DEMMEL and M. GU, *Global minimization of rational functions and the nearest GCDs*, J. Global Optim. **40(4)** (2008), 697–718.

[100] M.-T. NODA and T. SASAKI, *Approximate GCD and its application to ill-conditioned algebraic equations*, J. Comput. Appl. Math. 38(1-3) (1991), 335–351.

[101] A. M. OSTROWSKI, "Solution of Equations and Systems of Equations", 2nd ed., Pure and Applied Math., Academic Press, New York, 1966.

[102] V. Y. PAN, *On computations with dense structured matrices*, Math. of Comput. **55(191)** (1990), 179–190.

[103] V. Y. PAN, "Numerical Computation of a Polynomial GCD and Extensions", Research Report 2969, INRIA, Sophia-Antipolis, 1996.

[104] V. Y. PAN, *Approximate polynomial GCDs, Padé approximation, polynomial zeros and bipartite graphs*, Proceedings of the Ninth AMS-SIAM Symposium on Discrete Algorithms, ACM Press, New York, 1998, pp. 68-77.

[105] V. Y. PAN, *Computation of approximate polynomial GCD and an extension*, Information and Computation **167(2)** (2001), 71–85.

[106] V. Y. PAN, "Structured Matrices and Polynomials: Unified Superfast Algorithms", Birkhäuser, Boston, 2001.

[107] H. PARK, L. ZHANG and J. B. ROSEN, *Low-rank approximation of a Hankel matrix by structured total least norm*, BIT Numerical Mathematics **35(4)** (1999), 757–779.

[108] S. PILLAI and B. LIANG, *Blind image deconvolution using GCD approach*, IEEE Trans. Image Processing **8** (1999), 202–219.

[109] W. QIU, Y. HUA and K. ABED-MERAIM, *A Subspace method for the computation of the GCD of polynomials*, Automatica **33** (1997), 741–743.

[110] L. B. RALL, *Convergence of the Newton Process to Multiple Solutions*, Num. Math. **9** (1966), 23–37.

[111] G. RODRIGUEZ, *Fast Solution of Toeplitz- and Cauchy-Like Least-Squares Problems*, SIAM J. Matrix Anal. Appl. **28(3)** (2006), 724–748.

[112] J. B. ROSEN, *The gradient projection method for nonlinear programmin. II. Non linear constraints*, J. Soc. Indust. Appl. Math. **9** (1961), 514–532.

[113] S. M. RUMP, *Structured perturbations Part I: Normwise distances*, SIAM J. Matrix Anal. Appl. **25(1)** (2003), 1–30.

[114] D. RUPPRECHT, *An algorithm for computing certified approximate GCD of n univariate polynomials*, J. Pure and Appl. Alg. **139(1-3)** (1999), 255–284.

[115] M. SANUKI and T. SASAKI, "Computing Approximate GCD in Ill-Conditioned Cases", Proc. of the 2007 International Workshop on Symbolic-Numeric Computation, ACM, New York, 2007.

[116] T. SASAKI and M. SASAKI, *Polynomial Remainder Sequence and Approximate GCD*, ACM SIGSAM Bull. **31** (1997), 4–10.

[117] A. SCHÖNHAGE, "The Fundamental Theorem of Algebra in Terms of Computational Complexity", Mathematics Department, University of Tübingen, Germany, 1982.

[118] A. SCHÖNHAGE, *Quasi-GCD Computations*, J. Complexity, **1(1)** (1985), 118–137.

[119] I. SCHUR, *Über Potenzreihen die im Inneren des Einheitskreises beschränkt sind*, J. Reine Angew. Math. **147** (1917), 205–232. English translation in Oper. Theory Adv. Appl., Vol. 18, I. Gohberg (ed.), Birkhäuser, Boston, 1986, pp. 31–88.

[120] T. W. SEDERBERG and G.-Z. CHANG, *Best linear common divisors for approximate degree reduction*, Computer-Aided Design **25** (1993), 163–168.

[121] H. J. STETTER, "Numerical Polynomial Algebra", SIAM, Philadelphia, PA, 2004.

[122] G. W. STEWART, *On the Perturbation of Pseudo-Inverses, Projections and Linear Least Squares Problems*, SIAM Review 19(4) (1977), 634–662.

[123] M. STEWART, "Stable Pivoting for the Fast Factorization of Cauchy-Like Matrices", preprint, 1997.

[124] P. STOICA and T. SÖDERSTRÖM, *Common factor detection and estimation*, Automatica **33** (1997), 985–989.

[125] D. R. SWEET and R. P. BRENT, *Error analysis of a fast partial pivoting method for structured matrices*, In: "Adv. Signal Proc. Algorithms", Proc. of SPIE, T. Luk, ed., 2363 (1995), 266–280.

[126] A. TERUI, *An iterative method for computing approximate GCD of univariate polynomials*, Proceedings of the 2009 International Symposium on Symbolic and Algebraic Computation, ACM Press, 2009, 351–358.

[127] L. N. TREFETHEN and K.-C. TOH, *Pseudozeros of polynomials and pseudospectra of companion matrices*, Numer. Math. **68(3)** (1994), 403–425.

[128] R. VANDEBRIL, "SemiSeparable Matrices and the Symmetric Eigenvalue Problem", Ph.D. thesis, Katholieke Universiteit Leuven, 2004.

[129] R. VANDEBRIL, M. VAN BAREL and N. MASTRONARDI, "Matrix Computations and Semiseparable Matrices", Vol. 1, Linear systems, Johns Hopkins University Press, Baltimore, MD, 2008.

[130] R. VANDEBRIL, M. VAN BAREL and N. MASTRONARDI, "Matrix Computations and Semiseparable Matrices", Vol. II, Eigenvalue and singular value methods, Johns Hopkins University Press, Baltimore, MD, 2008.

[131] Y. WANG, "Computing Dynamic Output Feedback Laws with Pieri Homotopies on a Parallel Computer", Ph.D. thesis, University of Illinois, Chicago, 2005.

[132] J. R. WINKLER, *A companion matrix resultant for Bernstein polynomials*, Linear Algebra Appl. **362** (2003), 153–175.

[133] J. R. WINKLER and R. N. GOLDMAN, *The Sylvester Resultant Matrix for Bernstein Polynomials*, In: "Curve and Surface Design", Saint-Malo 2002, T. Lyche, M.-L. Mazure, L. L. Schumaker (eds.), Nashboro Press, Brentwood, Tennessee, 407–416, 2003.

[134] J. R. WINKLER and J. D. ALLAN, *Structured total least norm and approximate GCDs of inexact polynomials*, Journal of Computational and Applied Mathematics **215** (2008), 1–13.

[135] J. R. WINKLER and J. D. ALLAN, *Structured low rank approximations of the Sylvester resultant matrix for approximate GCDs of Bernstein polynomials*, Electronic Transactions on Numerical Analysis **31** (2008), 141–155.

[136] J. R. WINKLER and M. HASAN, *A non-linear structure preserving matrix method for the low rank approximation of the Sylvester resultant matrix*, Journal of Computational and Applied Mathematics **234** (2010), 3226–3242.

[137] J. R. WINKLER and X. LAO, *The calculation of the degree of an approximate greatest common divisor of two polynomials*, Journal of Computational and Applied Mathematics **235** (2011), 1587–1603.

[138] J.-C. YAKOUBSOHN, M. MASMOUDI, G. CHÈZE and D. AUROUX, *Approximate gcd à la Dedieu*, to appear in Applied Mathematics Electronic Notes.

[139] Z. H. YANG, *Polynomial Bezoutian matrix with respect to a general basis*, Linear Algebra Appl. **331** (2001), 165–179.

[140] C. J. ZAROWSKI, *The MDL criterion for rank determination via effective singular values*, IEEE Trans. Signal Processing **46** (1998), 1741–1744.

[141] C. J. ZAROWSKI, *QR-Factorization Method for Computing the Greatest Common Divisor of Polynomials with Inexact Coefficients*, IEEE Trans. Signal Processing **48** (2000), 3042–3051.

[142] T. Y. LI and Z. ZENG, *A rank-revealing method with updating, downdating, and applications*, SIAM J. Matrix Anal. Appl. **26(4)** (2005), 918–946.

[143] Z. ZENG, *Computing multiple roots of inexact polynomials*, Math. Comp. **74(250)** (2005), 869–903.

[144] Z. ZENG, *The approximate GCD of inexact polynomials Part I: a univariate algorithm*, to appear.

[145] Z. ZENG and B. H. DAYTON, "The Approximate GCD of Inexact Polynomials, Part II: a Multivariate Algorithm", Proceedings of the 2004 International Symposium on Symbolic and Algebraic Computation, ACM Press, New York, 2004.

[146] L. ZHI, *Displacement Structure in Computing the Approximate GCD of Univariate Polynomials*, Computer Mathematics, 288-298, Lecture Notes Series on Computing, Vol. 10, World Sci. Publ., River Edge, NJ, 2003.

[147] L. ZHI, K. LI and M.-T. NODA, "Approximate GCD of Multivariate Polynomials Using Hensel Lifting", MM Research Preprint 241-248, MMRC, AMSS, Academia Sinica, Beijing, 2001.

[148] L. ZHI and M.-T. NODA, "Approximate GCD of Multivariate Polynomials", Proc. ASCM 2000, World Scientific Press, 9-18, 2000.

[149] L. ZHI and Z. YANG, "Computing Approximate GCD of Univariate Polynomials by Structure Total Least Norm", MM Research Preprint 375-387, MMRC, AMSS, Academia Sinica, Beijing, 2004.

[150] R. ZIPPEL, "Effective Polynomial Computation", Kluwer Academic Publishers, Boston, 1993.

Index

THESES

This series gathers a selection of outstanding Ph.D. theses defended at the Scuola Normale Superiore since 1992.

Published volumes

1. F. COSTANTINO, *Shadows and Branched Shadows of 3 and 4-Manifolds*, 2005. ISBN 88-7642-154-8

2. S. FRANCAVIGLIA, *Hyperbolicity Equations for Cusped 3-Manifolds and Volume-Rigidity of Representations*, 2005. ISBN 88-7642-167-x

3. E. SINIBALDI, *Implicit Preconditioned Numerical Schemes for the Simulation of Three-Dimensional Barotropic Flows*, 2007. ISBN 978-88-7642-310-9

4. F. SANTAMBROGIO, *Variational Problems in Transport Theory with Mass Concentration*, 2007. ISBN 978-88-7642-312-3

5. M. R. BAKHTIARI, *Quantum Gases in Quasi-One-Dimensional Arrays*, 2007. ISBN 978-88-7642-319-2

6. T. SERVI, *On the First-Order Theory of Real Exponentiation*, 2008. ISBN 978-88-7642-325-3

7. D. VITTONE, *Submanifolds in Carnot Groups*, 2008. ISBN 978-88-7642-327-7

8. A. FIGALLI, *Optimal Transportation and Action-Minimizing Measures*, 2008. ISBN 978-88-7642-330-7

9. A. SARACCO, *Extension Problems in Complex and CR-Geometry*, 2008. ISBN 978-88-7642-338-3

10. L. MANCA, *Kolmogorov Operators in Spaces of Continuous Functions and Equations for Measures*, 2008. ISBN 978-88-7642-336-9

11. M. LELLI, *Solution Structure and Solution Dynamics in Chiral Ytterbium(III) Complexes*, 2009. ISBN 978-88-7642-349-9

12. G. CRIPPA, *The Flow Associated to Weakly Differentiable Vector Fields*, 2009. ISBN 978-88-7642-340-6

13. F. CALLEGARO, *Cohomology of Finite and Affine Type Artin Groups over Abelian Representations*, 2009. ISBN 978-88-7642-345-1

14. G. DELLA SALA, *Geometric Properties of Non-compact C R Manifolds*, 2009. ISBN 978-88-7642-348-2

15. P. BOITO, *Structured Matrix Based Methods for Approximate Polynomial GCD*, 2011. ISBN: 978-88-7642-380-2; e-ISBN: 978-88-7642-381-9

Volumes published earlier

H. Y. FUJITA, *Equations de Navier-Stokes stochastiques non homogènes et applications*, 1992.

G. GAMBERINI, *The minimal supersymmetric standard model and its phenomenological implications*, 1993. ISBN 978-88-7642-274-4

C. DE FABRITIIS, *Actions of Holomorphic Maps on Spaces of Holomorphic Functions*, 1994. ISBN 978-88-7642-275-1

C. PETRONIO, *Standard Spines and 3-Manifolds*, 1995. ISBN 978-88-7642-256-0

I. DAMIANI, *Untwisted Affine Quantum Algebras: the Highest Coefficient of* det H_η *and the Center at Odd Roots of 1*, 1996. ISBN 978-88-7642-285-0

M. MANETTI, *Degenerations of Algebraic Surfaces and Applications to Moduli Problems*, 1996. ISBN 978-88-7642-277-5

F. CEI, *Search for Neutrinos from Stellar Gravitational Collapse with the MACRO Experiment at Gran Sasso*, 1996. ISBN 978-88-7642-284-3

A. SHLAPUNOV, *Green's Integrals and Their Applications to Elliptic Systems*, 1996. ISBN 978-88-7642-270-6

R. TAURASO, *Periodic Points for Expanding Maps and for Their Extensions*, 1996. ISBN 978-88-7642-271-3

Y. BOZZI, *A study on the activity-dependent expression of neurotrophic factors in the rat visual system*, 1997. ISBN 978-88-7642-272-0

M. L. CHIOFALO, *Screening effects in bipolaron theory and high-temperature superconductivity*, 1997. ISBN 978-88-7642-279-9

D. M. CARLUCCI, *On Spin Glass Theory Beyond Mean Field*, 1998. ISBN 978-88-7642-276-8

G. LENZI, *The MU-calculus and the Hierarchy Problem*, 1998. ISBN 978-88-7642-283-6

R. SCOGNAMILLO, *Principal G-bundles and abelian varieties: the Hitchin system*, 1998. ISBN 978-88-7642-281-2

G. ASCOLI, *Biochemical and spectroscopic characterization of CP20, a protein involved in synaptic plasticity mechanism*, 1998. ISBN 978-88-7642-273-7